APPLIED CALC
USING MAPLE V

LARRY J. GOLDSTEIN

CALCULUS
&ITS APPLICATIONS
SEVENTH EDITION

BRIEF CALCULUS
AND ITS APPLICATIONS
SEVENTH EDITION

GOLDSTEIN · LAY · SCHNEIDER

PRENTICE HALL, Upper Saddle River, NJ 07458

Production Editor: *Tina Trautz*
Production Supervisor: *Joan Eurell*
Supplement Acquisitions Editor: *Audra Walsh*

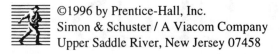
©1996 by Prentice-Hall, Inc.
Simon & Schuster / A Viacom Company
Upper Saddle River, New Jersey 07458

All rights reserved. No part of this book may be
reproduced, in any form or by any means,
without permission in writing from the publisher.

Printed in the United States of America

10 9 8 7 6 5 4 3 2 1

ISBN 0-13-375213-5

Prentice-Hall International (UK) Limited, *London*
Prentice-Hall of Australia Pty. Limited, *Sydney*
Prentice-Hall Canada Inc., *Toronto*
Prentice-Hall Hispanoamericana, S.A., *Mexico*
Prentice-Hall of India Private Limited, *New Delhi*
Prentice-Hall of Japan, Inc., *Tokyo*
Simon & Schuster Asia Pte. Ltd., *Singapore*
Editora Prentice-Hall do Brasil, Ltda., *Rio de Janeiro*

Contents

ABOUT THIS MANUAL vii

1 MAPLE V Basics 1
   1.1 A Typical MAPLE V Session 1
   1.2 Elementary Arithmetic in MAPLE 3
   1.3 Variables and Expressions 6
   1.4 Algebraic Manipulations 9
   1.5 Solving Equations 11
   1.6 Graphing 13

2 Functions 15
   2.1 Defining and Evaluating Functions 15
   2.2 Compound Interest 16
   2.3 Graphing Elementary Functions 17
   2.4 Translating Graphs of Functions 18
   2.5 Determining Zeros of Functions, Graphical Approach 19
   2.6 Determining Zeros of Functions, Analytic Approach 20
   2.7 Determining Intersection Points of Graphs, Graphical Approach 20
   2.8 Determining Intersection Points of Graphs, Analytic Approach 21
   2.9 Analysis of a Cost Function 22
   2.10 Functions Defined by Multiple Expressions . . . 22

3 The Derivative 24
   3.1 Exploring the Definition of the Derivative 24
   3.2 Tangent Lines and the Derivative 25
   3.3 Calculating Derivatives Symbolically 26
   3.4 Approximation of a Curve by Its Tangent Line . 27
   3.5 Calculating Limits 29
   3.6 Itemized Deductions on Tax Returns 30
   3.7 Calculating Limits as x Approaches ∞ or $-\infty$ . . 31
   3.8 The Rate of Change of Baseball Salaries 32
   3.9 Analysis of a Falling Ball 32
   3.10 Analysis of a Psychology Experiment 33

4 Applications of the Derivative 35
- 4.1 Describing Curves 35
- 4.2 Asymptotes of Curves 36
- 4.3 Determining Relative Maxima and Minima 37
- 4.4 Determining Concavity and Inflection Points ... 38
- 4.5 Analysis of a Medical Experiment 38
- 4.6 Analysis of a Botanical Study 39
- 4.7 Analysis of a Medical Experiment 40
- 4.8 Mathematical Model of an Illness 41
- 4.9 Analysis of Coffee Consumption 41
- 4.10 Analysis of the Sales of Cough and Cold Medicines 42

5 Techniques of Differentiation 44
- 5.1 The Rate of Change of the Election Function .. 44
- 5.2 Calculus and Technology Complement Each Other, I 45
- 5.3 Calculus and Technology Complement Each Other, II 45
- 5.4 Differentiation of Products and Quotients 46
- 5.5 The Chain Rule 47
- 5.6 Plotting Implicitly-Defined Functions 48

6 The Exponential and Natural Logarithm Functions .. 49
- 6.1 Calculating With Exponential Functions 49
- 6.2 Slopes of Exponential Functions 50
- 6.3 Graphs of Exponential Functions 51
- 6.4 Maxima and Minima Involving Exponential Functions 52
- 6.5 Normal Curves 53
- 6.6 Calculating With the Natural Logarithm Function 54
- 6.7 Graphs Involving the Natural Logarithm Function 55
- 6.8 Solving Equations Involving Exponential and Logarithmic Functions 55
- 6.9 Testing the Laws of Logarithms 56

7 Applications of the Exponential and Natural Logarithm Functions 57
- 7.1 Internal Rate of Return of an Investment 57

	7.2	Elasticity of Demand 57
	7.3	The Learning Curve 58
	7.4	The Yield Curve 59
	7.5	Spread of an Epidemic 60
	7.6	Analysis of the Effectiveness of an Insect Repellent 60
	7.7	Analysis of the Absorption of a Drug 61
	7.8	Analysis of the Growth of a Tumor 62
	7.9	Growth of a Bacteria Culture With Growth Restrictions 62
8	**The Definite Integral** **64**	
	8.1	Determining Antiderivatives 64
	8.2	Determining Antiderivatives, II 65
	8.3	Antidifferentiation in Closed Form Is Not Always Possible 66
	8.4	Estimation of Integrals Using Riemann Sums . . 66
	8.5	Calculating Definite Integrals 68
	8.6	Exploring the Fundamental Theorem of Calculus, I 69
	8.7	Exploring the Fundamental Theorem of Calculus, II 69
	8.8	The Area Between Curves 70
	8.9	The Volume of a Solid of Revolution 71
	8.10	Consumers' Surplus 71
9	**Functions of Several Variables** **73**	
	9.1	Defining and Evaluating Functions of Several Variables 73
	9.2	Graphing Functions of Two Variables 74
	9.3	Graphing Level Curves 75
	9.4	Calculating Partial Derivatives 75
	9.5	Solving Optimization Problems in Two Variables 77
10	**The Trigonometric Functions** **78**	
	10.1	Evaluating Trigonometric Functions 78
	10.2	Graphing Trigonometric Functions 79
	10.3	Amplitude and Period of Sine and Cosine Functions 79
	10.4	Analysis of Blood Pressure Readings 80

10.5	The Lotka-Volterra Predator-Prey Model	81
10.6	Number of Hours of Daylight	82
10.7	Graphs of Functions Formed From Trigonometric Functions	83

11 Techniques of Integration ... 84

11.1	More Antidifferentiation	84
11.2	Still More Antidifferentiation	84
11.3	The Trapezoid Rule	85
11.4	Simpson's Rule	86
11.5	Comparing the Accuracy of the Trapezoid and Simpson's rules	88
11.6	Approximating Definite Integrals With Specified Accuracy	89
11.7	Calculating Improper Integrals	90

12 Calculus and Probability ... 91

12.1	The Chi-Square Probability Density Function	91
12.2	Calculating Expected Value and Variance	92

13 Taylor Polynomials and Infinite Series ... 93

13.1	Approximation by Taylor Polynomials	93
13.2	Bernoulli Numbers	94
13.3	The Newton-Raphson Method	94
13.4	Two Interesting Infinite Series	94

ABOUT THIS MANUAL

This manual provides technology-based material to supplement your textbook for the standard applied calculus course.

The manual is software-specific. That is, its discussions and command descriptions are specifically geared to using MAPLE V. For other symbolic programs or graphing calculators, see the other manuals of this series.

The first two chapters are devoted to descriptions of the basics of MAPLE algebraic manipulations and graphics. The remaining chapters correspond to the standard chapters in your applied calculus text. Each chapter provides a series of technology discussions followed by corresponding exercise sets. Some discussions describe how to do various calculus computations using MAPLE V. Other discussions center on applications problems which involve calculations which use the concepts in the chapter, but involve calculations which are too burdensome to be done manually.

This manual may be used with the full version of MAPLE V or with the Student Version of Maple, on either the IBM or Macintosh platforms.

MAPLE V Basics

In this chapter, we discuss how to perform basic arithmetic and algebraic manipulations using MAPLE V.

1.1 A Typical MAPLE V Session

Starting MAPLE V

To start Maple under MS Windows or on the Macintosh, open the group or folder containing the MAPLE V icon and double click on the icon.

To start MAPLE V under MS DOS, first either make current the directory containing the MAPLE V program or make sure that that directory is in your DOS PATH. Then at the DOS prompt, type:
`maple`
and press ENTER.

Maple Commands

Once the program starts, you will see the MAPLE V main screen display, which consists of menu items across the top of the screen and an area to enter commands.

In a MAPLE session, you type in a command and MAPLE executes it and displays the result, which may be a number, expression, list, graph, or any of the other data types which MAPLE is capable of handling. For example, consider the problem of adding 2 and 3. Type in the command:
`2+3;`

Applied Calculus Using MAPLE V

and press ENTER. MAPLE responds with the answer:

$$5$$

Note that MAPLE displays the answer in the center of the line and the command at the left side of the screen, so it is easy to differentiate between commands you enter and output from the program.

Note that a MAPLE command ends with a semicolon. If you attempt to type
`2+3`
and type ENTER, MAPLE will make no response. It is waiting for the semicolon to tell it that the command is complete. If you now type a semicolon, MAPLE will execute the command. Even though the semicolon appears on a separate line, MAPLE will interpret it as ending the most recent command sent to it. This method of interpretation allows commands to extend over as many lines as you wish. This can help in keeping input readable. But more about that later.

We will learn the proper form in which to enter MAPLE commands. If you enter a command that MAPLE doesn't understand, it will respond with the message:
`syntax error`
and will point to the place in the command where it first does not understand. At this point, you may reenter the command (in corrected form) or enter another command.

Ending a MAPLE V Session

To end a MAPLE session, type the command:
`quit`

1.2 Elementary Arithmetic in MAPLE

Arithmetic Operations

MAPLE V uses the following symbols for arithmetic operations:

Addition	+
Subtraction	−
Multiplication	*
Division	/
Power	^

Moreover, MAPLE observes the same order of operations as is standard in arithmetic and algebra.

For example, to compute the value of:

$$5 \times 3 - 2 \div 3^6$$

, we would use the MAPLE command:
5*3-2/3^6;
MAPLE responds with the answer:

$$\frac{10933}{729}$$

You may use parentheses in MAPLE just as you would in algebra. For example, the problem:

$$-3(7-8) + 3^{2-4}$$

would be input in the form:
-3*(7-8)+3^(2-4);
Note that it is necessary to add a set of parentheses around the exponent in order for MAPLE to interpret it correctly. MAPLE responds with the answer:

$$\frac{28}{9}$$

Note several important differences between MAPLE results and those obtained from a calculator. First, MAPLE outputs answers in fractional form, where appropriate. (A calculator converts all answers into decimal form.) Second, MAPLE does its calculations to any number of significant digits. For example, consider the command:

`3^64;`

MAPLE outputs the answer:

$$3433683820292512484657849089281$$

(MAPLE does have limits, but you are unlikely to reach them in the course of any computations you encounter.)

Parentheses

You may use parentheses just as they appear within mathematical expressions. Note, however, that you may not use brackets [ ] or braces { } as grouping symbols. (MAPLE reserves brackets and braces for special purposes.) If these appear in an expression to compute, you must key them in as parentheses.

Note also, that in keying in certain expressions, it is necessary to introduce parentheses which are not present in usual mathematical notation. We have seen an example above in the case of an expression in an exponent.

Similarly, in keying in a fraction whose numerator or denominator is an expression, you must use parentheses. For example, the fraction:

$$\frac{5+8}{3-1}$$

must be keyed in as:

`(5+8)/(3-1);`

Scientific Notation

You may enter numbers in scientific notation. For example, to enter the number

$$1.09 \times 10^{-9}$$

Chapter 1 – MAPLE V Basics

you would enter:
`1.09*10^(-9);`

You may mix decimal notation and scientific notation throughout a problem.

Decimal Equivalents and Numerical Precision

You may obtain the decimal form of an answer by including a decimal point in at least one number within the command. For example, to obtain the value of $1/3 + 2/7$ in decimal form, you may enter the problem using the command:
`1./3+2/7;`
Maple gives the result:

$$.6190476190$$

You may refer to the most recent answer using quotation marks ". For example, the command
`"+3;`
will cause MAPLE to display:

$$3.619047619$$

You may control the number of significant digits by using the command `Digits`. (Note the capital letter D. This is essential. MAPLE distinguishes between upper and lower case letters.) For example, to set the number of significant digits to 5, give the command:
`Digits := 5;`
In response to this command, MAPLE just repeats the command. However, in all subsequent commands the number of significant digits will be 5.

Another way of converting output into decimal form is to use the command `evalf`. For example, the command
`evalf(1/3+2/7,3);`
displays the value of $1/3 + 2/7$ to 3 significant digits, namely .619.

Exercises

Calculate the following. In case the answer is a fraction, determine its decimal value to 4 significant digits.

1. $\frac{15}{11} - \frac{1}{19}$
2. $\frac{11}{29^3} \cdot \frac{43}{\frac{9}{19} - \frac{2}{3}}$
3. $\left(\frac{3}{4}\right)^{10}$
4. 48^{11}
5. $5.1 - 3.7 + 1.8 \times 2.27$
6. 37% of $48,385.91$
7. $(1 + .06/12)^{18}$
8. $3 - 12.11^{8.2}$
9. $3 - 4 \cdot (7.9 - 1.1)$
10. The average of the numbers 5.5, 7.15, 3.19, 4.88, 6.11
11. $1/3 + 5/6$ Give the answer both as a decimal and a fraction.
12. $5 + \dfrac{3.102 - 4.01}{2.09}$
13. $1.001 - (1 - .01/12)^{300}$
14. $1/3^4 - 7$

1.3 Variables and Expressions

Variables

MAPLE allows you to use variables, which are just place holders for numbers. Variable names can be single letters, such as c or s. Note that

Chapter 1 - MAPLE V Basics

MAPLE makes a distinction between upper and lower case letters. So the variables S and s are considered different.

MAPLE allows you to use long variable names, such as `length` and `amount`.

It is not necessary to do anything special to define a variable. Just use it within a command.

To assign a value to a variable, use the symbol :=. For example, to assign the variable z the value 1.71, we would use the command:
`z := 1.71;`
Note that assignment is indicated by := and **not** just =.

To see the current value of a variable, just give a command consisting of the variable name. For instance, to see the current value of z, you may give the command:
`z;`

Once a variable has been assigned a value, it retains that value throughout the session, until that value is changed (using another assignment command, for example).

Variables do not necessarily need to be assigned a value. As in algebra, MAPLE can work with symbolic variables. (See below for a discussion of doing algebraic calculations using variables.) If you ask for the value of a symbolic variable, MAPLE displays nothing.

You may "unassign" a variable, that is, to return it to a symbolic state. For example, to unassign the variable z above, use the command:
`z := 'z';`

Expressions

You may enter any algebraic expression in MAPLE. For example, you may enter the polynomial $z^2 - 3z + 1$ by giving the command:
`z^2-3*z+1;`

Applied Calculus Using MAPLE V

Note that it is necessary to explicitly enter the multiplication $3z$ using the multiplication symbol. MAPLE needs this to tell where the number stops and the variable name begins. Similarly, if you wish to enter the product of two variables xy, you must explicitly state the product:

x*y

Otherwise, MAPLE would think that you are referring to a variable with name **xy**. When you enter the above expression MAPLE responds with the value of the expression if **z** has a value; it just displays the expression if **z** is a symbolic variable.

It is often useful to give an expression a name. This is helpful, for example, when you wish to evaluate the expression for several values of the variable. For example, to store the above expression under the name **q**, you may use the command

q := z^2-3*z+1;

MAPLE responds by repeating the command. However, if you give the command

q;

then MAPLE will give the current value of **q**. If you change the value of **z** and give the command again, the new value of **q** will be displayed.

In the above discussion, **q** is treated just as any other variable. For example, to unassign it, you would use the command

q := 'q';

You may use an expression within an expression. For example, you may use the command

r := q^2;

to assign **q** the expression **(z^2-3*z+1)^2**.

Exercises

Write commands to assign to

A

the following expressions. Determine the values of the expressions for **x** equal to the values 1.5, 10, $\frac{17}{3}$, $2^{1/3}$. After determining each answer in exact form, determine its value to 10 significant digits.

1. $x^3 - 2x$
2. $3x + 1$
3. $\dfrac{x^2}{x^2 + 1}$
4. $\sqrt{x+1}$
5. $(x-1)^{10}$

1.4 Algebraic Manipulations

MAPLE is able to perform all of the standard algebraic operations on expressions. In this section, we will introduce you to these capabilities.

The command **expand** performs multiplications and combines like terms. For example, the command
`expand((x+1)*(x-2));`
gives the answer:

$$x^2 - x - 2$$

Expand is quite powerful. For example, you should try to use it to calculate the binomial expansion of $(x+y)^{50}$. See how quickly MAPLE can perform a symbolic manipulation that is barely possible by hand.

The command **simplify** performs simplification of algebraic expressions. For example, the command
`simplify(1/x+1/(x-1));`
gives the algebraic sum:

$$\frac{1}{x} + \frac{1}{x-1}$$

namely,

$$\frac{2x-1}{x(x-1)}$$

In adding algebraic fractions, MAPLE gives the answer in lowest terms. Note, however, that it does not automatically give you any values of the variable which are excluded since they are zeros of the denominator. (We will show you how to determine the values in the next section.)

You can determine the numerator and denominator of an expression using the commands **numer** and **denom**. For example, to determine the numerator and denominator of the last expression and assign them to the variables n and d, we would use the commands:

```
q := ";
n := numer(q);
d := denom(q);
```
In response to these commands, n is assigned the value $2x - 1$ and d the value $x(x - 1)$.

The command **simplify** also uses the laws of exponents, where appropriate to simplify an expression. For example, the command:
```
simplify(x^(3/4)*x^(1/2));
```
gives the result: $x^{5/4}$

The command **factor** may be used to factor an expression. For example, the command:
```
factor(x^11-81*x;)
```
gives the result: $x(x^5 - 9)(x^5 + 9)$

Exercises

Use MAPLE to perform the following algebraic operations.

1. $(2x - 1)^7$
2. $(3x - 9)(x + 4)$
3. $(x - 2)(2x + 4)(x - 3)(3x + 11)$
4. $\dfrac{x^2 - 1}{x^3 + 1} + \dfrac{1}{x - 1}$
5. $\dfrac{3x^2 - 2x + 1}{x^3 + 2x - 1} + x^3 + \dfrac{1}{x + 2}$
6. $(x - 1)^9 \left(\dfrac{x + 1}{x^2 - 1}\right)^{12}$

1.5 Solving Equations

Equations in One Variable

You may solve equations using the `solve` command. For example, to solve the equation

$$x^2 - 2 = 0$$

you may give the command:
`solve(x^2-2=0);`
Note that an equation is stated using the symbol = rather than the symbol := used in making assignments. In response to the above command, MAPLE gives the answer:

$$\sqrt{2}, -\sqrt{2}$$

Two points about this answer are worthy of note. First, the answer is exact. This is true in general about the command `solve`, namely that it looks for solutions in exact form. Second, note that the answer is expressed in terms of the square root. MAPLE will use square roots wherever they occur. However, other non-integer, rational powers (e.g. $\frac{1}{3}$, $-\frac{17}{5}$ are displayed using fractional exponents.

The `solve` command may be used to solve an equation in two variables for one variable in terms of the other. For example, to solve the equation

$$xy - 3 = x$$

for y in terms of x, you may use the command
`solve(x*y-3=x,y);`
The last `y` in the parentheses indicates that MAPLE should solve for `y`. The answer given is:

$$-\frac{-3-x}{x}$$

(This may not quite be in the form you expect, but that often happens when you solve problems using a symbolic math program.)

Numerical Approximation to Solutions

You may numerically approximate the solutions of an equation using the command `fsolve`. For example, to solve the equation $x^2 - 2 = 0$ numerically, use the command

`fsolve(x^2-2=0);`

MAPLE will display the real roots of the equation using the current number of digits accuracy specified. Assume that 5 digits is currently called for, MAPLE displays:

$$-1.4142, 1.4142$$

Systems of Equations

You may use `solve` and `fsolve` to solve systems of equations. When dealing with more than one equation, you must enclose the equations within braces, with individual equations separated by commas. Moreover, you must include a list of the variables to solve for, enclosed within braces. For example, to solve the system of linear equations:

$$\begin{cases} x + y = 3 \\ x - y = 7 \end{cases}$$

you would use the command:

`solve({x+y=3,x-y=7},{x,y});`

MAPLE responds with the answer"

$$\{y = -2, x = 5\}$$

The format for using `fsolve` to obtain numerical rather than exact solutions is similar.

Exercises

Solve the following equations or systems. First determine the solutions exactly. Then determine the solutions to 5 significant digits.

1. $x^3 = 5$
2. $3x - 7(x + 3) = 19$

Chapter 1 – MAPLE V Basics

3. $x^2 - 12x + 14 = 0$
4. $x^3 - 2x^3 + x - 5 = 0$
5. $xy - 5y = 12x^2$ (for y)
6. $A = 4\pi r^2 + 2\pi r$ (for r) Note that you may enter the number π in MAPLE by typing `Pi;`.
7. $x - 3x = 7, \; 2x + 8y = 11$
8. $xy^2 = 11, \; x + y = 1$

1.6 Graphing

MAPLE has extensive facilities for graphing equations. For graphing equations of the form y = [expression in x], use the `plot` command. For example, to plot the graph of $y = 3x - 1$ $(-2 \leq x \leq 2)$, you may use the command
`plot(3*x-1, x=-2..2);`
MAPLE will automatically determine the range of y-values needed and scale the graph accordingly. You may explicitly state the range of y-value allowed. For example, if you wish to restrict the y-values on the graph to lie between 0 and 10, the above command may be modified to read:
`plot(3*x-1, x=-2..2, y=0..10);`

If you wish to plot more than one equation on the same graph, include the expressions for y in a list enclosed in braces, with expressions separated by commas. For example, if you wish to add to the last plot the graph of $y = x^2$ (with the same restrictions on x and y), you may use the command:
`plot({3*x-1,x^2}, x=-2..2, y=0..10);`

For graphs which are not expressed in the form y = [expression in x], you may use the command `implicitplot`. (This command may actually be used on all equations, but is much slower than `plot`.) For example, to plot the graph of

$$x^2 + y^2 = 1, \quad -1 \leq x \leq 1, \quad -1 \leq y \leq 1$$

,you may use a command of the form:
`with(plots): implicitplot(x^2+y^2=1,x=-1..1,y=-1..1);`
MAPLE will display the graph which is a circle of radius 1 with center at the origin.

Exercises

Perform the following graphing operations.

1. Display the graph of $Y = 4 - X^2, -3 \leq X \leq 3$.
2. Examine the graph of 1. and redisplay the graph so that only the portion of the graph having positive Y-coordinates is displayed.
3. Examine the graph of 2. and redisplay the graph so that the graph fills the display vertically.
4. Add to the graph of 3. the graph of $Y = X + 1$.
5. Redisplay the graph of 4. omitting the graph of $Y = 4 - X^2$.
6. Graph the equation: $y = x^3 - 2x^2 + 1$. Use the graph to estimate the x-intercepts of the graph accurate to 0.1. Confirm these estimates by solving numerically for the zeros of the right side.
7. Graph the equation

$$xy = 1, \quad -2 \leq x \leq 2, \quad -10 \leq y \leq 10$$

   Describe what happens to the graph for values of x near 0.

8. Graph the equation:

$$4x^2 + 3y^2 = 1$$

   By experimentation, determine suitable ranges for x and y.

Functions

In this chapter, we apply the various capabilities of the MAPLE to study functions.

2.1 Defining and Evaluating Functions

In the preceding chapter, we showed how to assign names to expressions. Let us now go a step further and show how to define functions within MAPLE. Recall that a function of **x** is a rule which assigns to a value of **x** a number, called the *function value* at **x**. A function may be assigned any variable name. As an example, suppose we wish to define the function $f(x) = x^2$. The function name in this case would be **f**. To define the function, we would give the MAPLE command:

```
f := x -> x^2;
```

Note that the symbol -> consists of the hyphen - followed by the inequality symbol >.

Once a function is defined, you may use it throughout the MAPLE session.

You may evaluate a function at a value of the variable by inputting the function name followed by the value in parentheses. For example, to obtain the function value $f(2)$ for the function defined above, you would give the MAPLE command:

```
f(2);
```

MAPLE returns the answer:

$$4$$

You may use **f(x)** to obtain the expression for **f**. For example, the MAPLE command:

```
f(x)*x^3;
```

returns the expression:

$$x^5$$

You may plot functions using the `plot` command. For example, to plot the function f above, for $-3 \leq x \leq 3$, you would use the command:
`plot(f(x),x=-3..3);`

Exercises

For each of the following functions:

a. Determine $f(a)$ for the value of a indicated.

b. Plot f for X in the range $-10 \leq x \leq 10$.

c. By examining the graph, determine the range of values in $[-10, 10]$ for which f is defined.

1. $f(X) = X^3$, $a = 5.98$
2. $f(X) = 3X^2 - 8X + 1$, $a = -9.11$
3. $f(X) = 1/(X+1)$, $a = -1.0798 \times 10^5$
4. $f(X) = \sqrt{X^2 - 1}$, $a = -3.22$
5. $f(X) = \dfrac{X-2}{X}$, $a = 3.02793$
6. $f(X) = X(X^4 - 1)^3$, $a = 1.01$

2.2 Compound Interest

Suppose that P dollars are deposited in a savings account paying annual interest at rate r compounded k times per year. Then the amount A in the account after X years is given by the formula:

$$A = P\left(1 + \frac{r}{k}\right)^{kX}$$

Chapter 2 – Functions

Exercises

Suppose that $500 is deposited into an account paying 6% interest compounded daily. (Assume that a year has 360 days and that each quarter has 90 days.) Solve the following problems:

1. Determine the amount in the account at the end of each quarter for the first 10 quarters.
2. How much money is in the account at the end of 6.75 years?
3. How much money is in the account at the end of 10 years?
4. How many quarters before there is more than $1300 in the account?
5. How many quarters before there is more than $5000 in the account?
6. How much money is in the account after 7 months?
7. How much money is in the account after 2 years 5 months?

2.3 Graphing Elementary Functions

This project is designed to acquaint you with the graphs of some elementary functions.

Exercises

Graph the following functions.

1. Linear functions: $f(x) = ax + b$ for $a = 1, b = 3$; $a = -2, b = 4$; $a = 0, b = 1$; $a = 4, b = -7$.
    (a) Describe the graph of a linear function.
    (b) What is the geometric significance of the coefficient b?
    (c) Experiment with other positive and negative values of a to determine what happens to the graph as the value of a increases? Decreases?

2. Quadratic functions: $f(x) = ax^2 + bx$, for $a = 1, b = -1$; $a = 2, b = 0$; $a = -1, b = 3$; $a = 4, b = -2$.

   (a) Describe the graph of a quadratic function.
   (b) What is the significance of the coefficient a to the graph?

3. Cubic functions: $f(x) = x^3 - ax$ for $a = -2, -1, -.5, 0, .5, 1, 2$. What is the domain of each of these functions?

4. Graph $x + \dfrac{1}{x}$ for $.5 \leq x \leq 10$. Graphically determine the domain and range of this function.

5. Graph $f(x) = \sqrt{2-x}$. Graphically determine the domain and range of this function.

2.4 Translating Graphs of Functions

In this set of explorations, you are asked to discover the relationship between the graph of $f(X)$ and the graphs of $f(X + c)$ and $f(X) + c$.

Exercises

Let $f(x) = x^2$.

1. Graph the functions $f(x+1)$, $f(x-1)$, $f(x+2)$, $f(x-2)$ on a single coordinate system.

2. On the basis of the evidence accumulated in 1., make a guess about the relationship between the graph of a general function $f(x)$ and the graph of $f(x+h)$.

3. Test your guess in 3. using the functions $g(x) = x^3$ and $f(x) = \sqrt{x}$ and various values of h.

4. Graph the functions $f(x) + 1$, $f(x) - 1$, $f(x) + 2$, $f(x) - 2$.

5. On the basis of the evidence accumulated in 4., make a guess about the relationship between the graph of a general function $f(x)$ and the graph of $f(x) + h$.

6. Test your guess in 5. using the functions $g(x) = x^3$ and $f(x) = \sqrt{x}$ and various values of h.

7. Using what you have learned above, sketch the graph of $f(x) = (x-1)^2 + 2$ without using a graphing calculator. Check your result using MAPLE.

2.5 Determining Zeros of Functions, Graphical Approach

By examining the graph of a function, you may determine the approximate values of the zeros. By regraphing the function to more closely examine the graph near each of the zeros, you may obtain a more accurate approximation to the zeros. By repeated re-enlargements, you may obtain numerical approximations to the zeros to any desired degree of accuracy. Use this technique to determine the zeros of the following functions in the interval $-10 \leq x \leq 10$.

Exercises

Determine each zero to within an accuracy of 0.1.

1. $f(x) = x^2 - x - 2$
2. $f(x) = x^3 - 3x + 4$
3. $f(x) = \sqrt{x-2} - x = 2$
4. $f(x) = \dfrac{x}{x+1} - x^2 + 1$

2.6 Determining Zeros of Functions, Analytic Approach

You may determine the exact values of the zeros of a function using the `solve` command. For example, to determine the zeros of f, you would use the command:
`solve(f(x)=0);`
To obtain numerical approximation to the zeros, you would use a similar command with `solve` replaced by `fsolve`.

Exercises

Use the above approach to determine numerical approximations of the zeros of the following functions. Obtain your results to 5 significant digits. Compare the results with those obtained graphically in the preceding section.

1. $f(x) = x^2 - x - 2$
2. $f(x) = x^3 - 3x + 4$
3. $f(x) = \sqrt{x-2} - x = 2$
4. $f(x) = \dfrac{x}{x+1} - x^2 + 1$

2.7 Determining Intersection Points of Graphs, Graphical Approach

You may determine the intersection point of a pair of graphs by repeatedly enlarging the graph in the vicinity of the intersection point. In this way, you may obtain the coordinates of the intersection point to any desired degree of accuracy.

Exercises

Use the above graphical approach to approximate the coordinates of each intersection point of the graphs of the given functions to within .1.

1. $f(x) = 2x - 1$, $g(x) = x^2 - 2$
2. $f(x) = -x - 2$, $g(x) = -4x^2 + x + 1$
3. $f(x) = x^2 - 3x$, $g(x) = x^3 + 2x^2 - 4x + 1$
4. $f(x) = \dfrac{1}{x}$, $g(x) = \sqrt{x^2 - 1}$

2.8 Determining Intersection Points of Graphs, Analytic Approach

You may use `solve` and `fsolve` to determine the coordinates of the intersection points of the graphs of the functions `f` and `g`. The x-coordinate is just the solution of the equation

`f(x)=g(x)`

Once the value of `x` is found, you may determine the value of `y` as the function value `f(x)`.

Exercises

Apply the above approach to determine the intersection points of the following graphs. Compare the results with the graphical approach of the preceding section.

1. $f(x) = 2x - 1$, $g(x) = x^2 - 2$
2. $f(x) = -x - 2$, $g(x) = -4x^2 + x + 1$
3. $f(x) = x^2 - 3x$, $g(x) = x^3 + 2x^2 - 4x + 1$
4. $f(x) = \dfrac{1}{x}$, $g(x) = \sqrt{x^2 - 1}$

2.9 Analysis of a Cost Function

Suppose that the cost of producing X units per hour is given by the cost function $C(X) = 150 + 59X - 1.8X^2 + .02X^3$.

Exercises

Perform the following analysis on the mathematical model described above.

1. Graphically determine the y-intercept of the graph. What is the applied significance of the y-intercept?
2. Determine the cost of producing 115 units per hour.
3. By how much does the cost increase if production is increased from 115 to 116 units per hour?
4. Suppose that cash flow restricts operations to spending no more than $15,000 per day. Assuming that production is spread out over 10 hours each day, what will be the number of units produced each hour?

2.10 Functions Defined by Multiple Expressions

Many functions are defined using more than one expression. For example, consider the function:

$$f(x) = \begin{cases} x & \text{if } x \leq 1 \\ x^2 & \text{if } x > 1 \end{cases}$$

For values of x in the interval $(-\infty, 1]$, the function is defined by the formula x, whereas for x in the interval $(1, \infty)$, the function is defined by the formula x^2. MAPLE allows you to introduce such functions into your session using the **proc** command. Here **proc** is short for *procedure*. This

Chapter 2 – Functions

command allows you to specify a procedure for computing a function. The procedure can contain any number of steps. Here is how to define the function above:

```
f := proc(x) if x <= 1 then x else x^2 fi end;
```

To explain this procedure code, let's take it a small chunk at a time. The line `proc(x)` tells MAPLE that a procedure is being defined and that when the procedure is used, it is to make a computation using the value of x specified in `f(x)`. Using this value of x, the procedure makes a decision: If $x \leq 1$, then the procedure returns the value of f as x; if $x > 1$, then the procedure returns the value of f as x^2. Note that the symbol $\leq$ is written for computer usage as <=. Similarly, the symbol $\geq$ would be written as >=. The clause which begins `if` must be ended with the keyword `fi` (which is `if` backwards). The keyword `end` indicates the end of the procedure. You should type the statement on a single line. MAPLE allows you to type lines which extend beyond the width of the screen. Just keep typing, but don't press ENTER.

Exercises

For each of the following functions:

1. Determine the following function values: $f(1)$, $f(3)$, $f(-1)$.
2. Graph the function using the range $-10 \leq x \leq 10$.

1. $f(x)$ defined above.

2. $f(x) = \begin{cases} x & \text{if } x < 1 \\ 2x - 1 & \text{if } x \geq 1 \end{cases}$

3. $f(x) = \begin{cases} x & \text{if } x < 1 \\ x^2 & \text{if } x \geq 1 \end{cases}$

4. A multiple inequality, such as $1 \leq X \leq 2$ may be keyed in using two inequalities:

$$(1 <= X) * (X <= 2)$$

Use this fact to graph the function:

$$f(x) = \begin{cases} 3 & \text{if } 0 \leq x \leq 1 \\ 1 & \text{if } x > 1 \end{cases}$$

The Derivative

3.1 Exploring the Definition of the Derivative

Suppose that f is a function defined in an open interval containing $x = a$. The derivative $f'(a)$ is approximated by the difference quotient

$$\frac{f(a+h) - f(a)}{h}$$

where h assumes values approaching 0, such as 0.1, 0.01, 0.001, 0.0001,... . Such a sequence of values for h may be calculated as

$$h = 10\wedge(-n)$$

for $n = 1, 2, 3, 4, ...$ You can approximate $f'(a)$ by calculating the difference quotient

$$\frac{f(a+h) - f(a)}{h}$$

for $n = 1, 2, 3, 4, ...$

Exercises

For the following functions $f(x)$ and values $x = a$, calculate the value of the difference quotients for $n = 1, 2, ..., 10$. By examining the values produced, estimate the value of $f'(a)$.

1. $f(x) = x^2$, $a = 3$
2. $f(x) = x^2$, $a = -1$
3. $f(x) = \sqrt{x}$, $a = 4$
4. $f(x) = \sqrt{x}$, $a = \frac{1}{4}$

5. $f(x) = 1/x$, $a = 2$
6. $f(x) = 1/x$, $a = \frac{1}{2}$
7. $f(x) = (2x+1)^2(3x-1)$, $a = 1$
8. $f(x) = x^4 + 2x^2 - 1$, $a = 1$
9. $f(x) = \dfrac{x}{x+1}$, $a = 1$
10. $f(x) = x(2x-1)^8$, $a = 1$

3.2 Tangent Lines and the Derivative

You may use the `plot` command to draw graphs of functions along with tangent lines at particular points. For example, consider the function $f(x) = x^2$ which you may define with the command:

`f := x -> x^2;`

The tangent line to this function at $x = 3$ has slope

$$m = 2x = 2 \cdot 3 = 6$$

and so has equation:

$$y - 3^2 = 6(x - 3)$$
$$y = 6x - 9$$

So we may graph the curve and the tangent line using the command:

`plot({x^2,6*x-9},x=0..6);`

Note that we have chosen the `x`-range so that the point of tangency is in the center of the screen for easy viewing.

Exercises

For the following functions $f(x)$, draw the graph and the tangent line at $x = a$. Determine two points on the tangent line. Use this data to determine the value of $f'(a)$. Compare the value so determined with the corresponding value determined in the preceding exercise set.

1. $f(x) = x^2$, $a = 3$
2. $f(x) = x^2$, $a = -1$
3. $f(x) = \sqrt{x}$, $a = 4$
4. $f(x) = \sqrt{x}$, $a = \frac{1}{4}$
5. $f(x) = 1/x$, $a = 2$
6. $f(x) = 1/x$, $a = \frac{1}{2}$
7. $f(x) = (2x+1)^2(3x-1)$, $a = 1$
8. $f(x) = x^4 + 2x^2 - 1$, $a = 1$
9. $f(x) = \dfrac{x}{x+1}$, $a = 1$
10. $f(x) = x(2x-1)^8$, $a = 1$

3.3 Calculating Derivatives Symbolically

MAPLE has the ability to calculate derivatives of functions using the command `diff`. For example, to calculate the derivative of `x^2`, you would use the command:

`diff(x^2,x);`

Note that the `x` indicates the variable with respect to which you are differentiating.

To define the derivative as a function, you would use a command of the form:

`fprime := x -> diff(x^2,x);`

You could then use this function to calculate the derivative at `x=1.5` using the command.

`fprime(1.5);`

Chapter 3 -- The Derivative

Exercises

For each of the following functions:

    a.   Determine the derivative.

    b.   Determine the value of the derivative at the indicated value of a.

    c.   In each case, compare the derivative obtained in a. with the derivative obtained manually using the rules of differentiation.

1. $f(x) = x^2$, $a = 3$
2. $f(x) = x^2$, $a = -1$
3. $f(x) = \sqrt{x}$, $a = 4$
4. $f(x) = \sqrt{x}$, $a = \frac{1}{4}$
5. $f(x) = 1/x$, $a = 2$
6. $f(x) = 1/x$, $a = \frac{1}{2}$
7. $f(x) = (2x+1)^2(3x-1)$, $a = 1$ [Hint: For the exact value, multiply out the expression for the function.]
8. $f(x) = x^4 + 2x^2 - 1$, $a = 1$
9. $f(x) = \dfrac{x}{x+1}$, $a = 1$
10. $f(x) = x(2x-1)^8$, $a = 1$

3.4 Approximation of a Curve by Its Tangent Line

The tangent line to a curve at a point provides an approximation to the curve in the vicinity of the point. By examining a plot containing both the curve and the tangent line near the point of tangency, you may obtain an estimate of the maximum difference in height between the tangent line and the curve for x in a specific interval.

Applied Calculus Using MAPLE V

Exercises

For the following functions $f(x)$, do the following:

    a.   Draw the graph and its tangent line at $x = a$ for x in the range $a - 0.1 \leq x \leq a + .1$.

    b.   Determine the maximum deviation of the graph from its tangent line in the region of b.

1. $f(x) = x^2$, $a = 3$
2. $f(x) = x^2$, $a = -1$
3. $f(x) = \sqrt{x}$, $a = 4$
4. $f(x) = \sqrt{x}$, $a = \frac{1}{4}$
5. $f(x) = 1/x$, $a = 2$
6. $f(x) = 1/x$, $a = \frac{1}{2}$
7. $f(x) = (2x+1)^2(3x-1)$, $a = 1$
8. $f(x) = x^4 + 2x^2 - 1$, $a = 1$
9. $f(x) = \dfrac{x}{x+1}$, $a = 1$
10. $f(x) = x(2x-1)^8$, $a = 1$

Let $f(x) = 2x^2$.

11. Display the graphs of f and its tangent line at the point $(1, 2)$ on a single screen.

12. Examine the two graphs in the interval $.5 \leq X \leq 1.5$. Observe how the portion of the graph of f is almost straight.

13. By how much does the function graph differ from its tangent line over the interval?

14. Repeat 12. and 13. using the interval $.9 \leq X \leq 1.1$.

15. Repeat 12. and 13. using the interval $.99 \leq X \leq 1.01$.

3.5 Calculating Limits

MAPLE has a built-in capability to calculate limits. To calculate the limit:
$$\lim_{x \to a} f(x)$$
use the command:
`limit(f(x),x=a);`
(assuming that the function `f` has been previously defined). In response to this command MAPLE returns the value of the limit, provided that the limit exists and the error message `cannot evaluate boolean` if the limit does not exist.

The corresponding right and left hand limits
$$\lim_{x \to a^+} f(x)$$
$$\lim_{x \to a^-} f(x)$$

may be evaluated using the commands:
`limit(f(x),x=a, right);`
`limit(f(x),x=a, left);`

respectively.

Exercises

Calculate the following limits using MAPLE.

1. $\lim\limits_{x \to 1} \dfrac{x^2 - 1}{x - 1}$

2. $\lim\limits_{x \to 9} \dfrac{\sqrt{x^2 - 5x - 36}}{8 - 3x}$

3. $\lim\limits_{x \to 1} (1 - 4x)$

4. $\lim\limits_{x \to 2} (x^2 - 3x + 2)$

5. $\lim\limits_{x \to 2} \dfrac{x + 2}{x - 2}$

6. $\lim_{x \to 1} \dfrac{x-1}{x}$

7. $\lim_{x \to 1} \dfrac{-2}{\sqrt{x+16}+7}$

8. $\lim_{x \to 5} \dfrac{3x-15}{x^2-25}$

9. $\lim_{x \to 1^+} \dfrac{x-1}{\sqrt{x-1}}$

10. $\lim_{x \to 0^+} \dfrac{1}{x}$

3.6 Itemized Deductions on Tax Returns

The Internal Revenue Service code allows taxpayers to take deductions from their income for certain expenses, such as mortgage interest, medical expenditures and charitable contributions. These deductions are called *itemized deductions*. Typically, the amount of itemized deductions claimed rises with income.

Let y denote the average amount claimed for itemized deductions y on a tax return reporting x dollars of income. According to Internal Revenue Service data, y is a linear function of x. Moreover, in a recent year, income tax returns reporting $20,000 of income averaged $729 in itemized deductions, while returns reporting $50,000 averaged $1380.

Exercises

Solve the following exercises concerning itemized deductions.

1. Determine y as a function of x.
2. Graph the function of Exercise 1.
3. Graphically determine the slope of the line displayed.
4. Give an interpretation of the slope in applied terms.

5. Determine graphically the average amount of itemized deductions on a return reporting $75,000.

6. Determine graphically the income level at which the average itemized deductions are $5,000.

7. Suppose that the income level increases by $15,000. By how much do the average itemized deductions increase?

3.7 Calculating Limits as x Approaches ∞ or −∞

In MAPLE you may enter ∞ and $-\infty$ in commands using `infinity` and `-infinity`. Using these MAPLE constants, you may evaluate limits of the form:

$$\lim_{x \to \infty} f(x), \quad \lim_{x \to -\infty} f(x)$$

Exercises

Use MAPLE to determine the values of the following limits.

1. $\lim\limits_{x \to \infty} \dfrac{x^2 - 2x + 3}{2x^2 - 1}$

2. $\lim\limits_{x \to \infty} \left[\sqrt{x} - \sqrt{25 + x} \right]$

3. $\lim\limits_{x \to -\infty} 3 + \dfrac{1 + x}{x^2}$

4. $\lim\limits_{x \to \infty} \dfrac{-8x^2 + 1}{x^2 + 1}$

3.8 The Rate of Change of Baseball Salaries

Let y denote the average salary of a baseball player (in thousands of dollars) in year x (where years are measured with 1982 corresponding to 0). Throughout the 1980's and early 1990's (before the baseball strike of 1994-95), baseball salaries increased steeply. A mathematical model for these salaries is

$$y = 246 + 64x - 8.9x^2 + 0.95x^3$$

Exercises

Perform the following analysis of the above mathematical model.

1. Graph y as a function of x.
2. When was the average salary $300,000?
3. What was the average salary in 1990?
4. By how much did the average salary increase from 1990 to 1991?
5. Compute the derivative y'.
6. Use the derivative to estimate the amount that the average salary increased from 1990 to 1991.
7. Compare the answers derived in 4. and 6.

3.9 Analysis of a Falling Ball

The laws of physics imply that if a ball is thrown with initial velocity v from a height h, then the height of the ball t seconds after being thrown is given by the model

$$s(t) = \frac{1}{2}gt^2 + vt + h$$

where g is a physical constant, which is approximately equal to -32 ft/sec^2.

Suppose that a ball is thrown with an initial velocity 102 ft/sec from ground level. Then the above model gives the height t seconds after the ball is thrown as:

$$s(t) = 102t - 16t^2$$

Exercises

Perform the following analysis of the above mathematical model of a falling ball.

1. Use a graphing calculator to display the graphs of $s(t)$, $s'(t)$, $s''(t)$ for $0 \leq t \leq 10$. Use these graphs to answer the remaining questions.
2. How high is the ball after 2 seconds?
3. When, during descent, is the height 110 feet?
4. What is the velocity after 6 seconds?
5. When is the velocity 70 feet per second?

3.10 Analysis of a Psychology Experiment

In a psychology experiment[1], people improved their ability to recognize common verbal and semantic information with practice. Their judgement time after t days of practice was $f(t) = .36 + .77(t - .5)^{-.36}$.

Exercises

Perform the following analysis of the above mathematical model of judgement time.

---

[1] Anderson, John R. *Automaticity and the ACT Theory*, American Journal of Psychology, Summer 1992, vol. 105, No. 2, pp. 165–180.

1. Display the graphs of $f(t)$, $f'(t)$, $f''(t)$ for $.5 \leq t \leq 10$. Use these graphs to answer the following questions.

2. What was the judgement time after 2 days of practice?

3. After how many days of practice was the judgement time about .8 seconds?

4. After 2 days of practice, at what rate was judgement time changing with respect to days of practice?

5. After how many days was judgement time changing at the rate of $-.08$ seconds per day of practice?

Applications of the Derivative

4.1 Describing Curves

You can use MAPLE to determine the intervals in which a function is increasing or decreasing. To do this, use the solve command to determine where the derivative is positive or negative. For example, consider the function $f(x) = x^2$ with derivative $f'(x) = 2x$. You may solve the inequality

$$f'(x) > 0$$

using the command:
`solve(2*x > 0);`
MAPLE represents the answer in the form
`{0<x}`

In solving inequalities, the interval $1 < x < 2$ is represented by MAPLE as:
`{1<x, x<2}`
Moreover, the pair or conditions $x < -2$ or $x > 1$ is represented by:
`{x<-2}, {1<x}`

Exercises

Use the technique described above to determine the intervals in which the graphs of the following functions increase or decrease.

    1. $f(x) = x^2 - 5x - 1$

2. $f(x) = -2x^2 + x + 1$

3. $f(x) = x^3 - 3x + 1$

4. $f(x) = 2x^3 - 7x + 1$

4.2 Asymptotes of Curves

The graph of $f(x) = 1/x$ has a vertical asymptote $x = 0$, corresponding to the point at which the denominator is 0.

MAPLE (and all plotting programs) do not do well drawing the graphs with asymptotes. Display the graph of the function above. Note that the asymptote is rendered as a spike and that the graph looks as if the function is defined for $x = 0$. Here's the reason. The program computes points on the graph at successive values of x corresponding to screen dots across the x-axis. As it computes each point, the program plots it on the display. If the function is not defined at a particular value of x, no point is plotted. However, in order to give the display the appearance of a smooth curve, MAPLE draws a line connecting consecutive points on the graph. In this case, the program connects a point slightly to the left of $x = 0$ with a point slightly to right of $x = 0$. No point is plotted for $x = 0$ since the function is undefined there. You must get used to interpreting spikes as asymptotic behavior.

When sketching a graph of a rational function using a symbolic math program, you can locate the asymptotes by determining the zeros of the denominator. For example, to determine the vertical asymptotes of the function

$$f(x) = \frac{x}{x^3 - 3x + 1}$$

you would use MAPLE to solve the equation $x^3 - 3x + 1 = 0$. The solutions give the vertical asymptotes.

Exercises

Determine the vertical asymptotes of the following functions. Display the graph of each function and confirm the position of the asymptotes.

1. $f(x) = \dfrac{x+2}{3x-1}$

2. $f(x) = \dfrac{x^2}{x^3 - x + 1}$

3. $f(x) = \dfrac{2x+1}{x^4 - 2}$

4. $f(x) = \dfrac{x^3 - 1}{x^4 - 2x^2 - 3}$

4.3 Determining Relative Maxima and Minima

We may use MAPLE to determine possible relative extreme points by solving the equation $f'(x) = 0$. We may then apply the second derivative test to determine whether the values of x so determined are relative maxima, relative minima, or can't tell. In the last case, we may graph the function to determine the nature of the point.

Exercises

Use MAPLE to determine all relative extreme points of the following functions. Classify them as relative maximum or relative minimum points.

1. $f(x) = x^3 - 6x^2 + 12x - 6$

2. $f(x) = 1 - 3x + 3x^2 - x^3$

3. $f(x) = \dfrac{1}{1 - 20x}$

4. $f(x) = \dfrac{1}{x^2 + 1}$

4.4 Determining Concavity and Inflection Points

The graph of $f(x)$ is concave up in an interval in which $f''(x) > 0$ for all x in the interval. The graph of $f(x)$ is concave down in an interval in which $f''(x) < 0$ for all x in the interval. At an inflection point, $f''(x) = 0$.

Given a function f, we may use MAPLE to determine $f''(x)$. Then, by graphing $f''(x)$, we may determine the intervals in which the graph of f is concave up or concave down as well as the location of any inflection points.

Exercises

Determine the intervals in which graphs of the following functions are concave up and concave down by displaying the function and its second derivative on the same graph. Use a graphical approach to determine the inflection points of the graph.

1. $f(x) = x^3 - 6x^2 + 12x - 6$
2. $f(x) = 1 - 3x + 3x^2 - x^3$
3. $f(x) = \dfrac{1}{1 - 3x}$
4. $f(x) = \dfrac{1}{x^2 + 1}$

4.5 Analysis of a Medical Experiment

In a medical experiment[2], the body weight of a baby rat in the control group after t days was $f(t) = 4.96 + .48t + .17t^2 - .0048t^3$ grams.

---

[2] Johnson, Wogenrich, Hsi, Skipper, and Greenberg, *Growth Retardation During the Suckling Period in Expanded Litters of Rates; Observations of Growth Patterns and Protein Turnover*, Growth, Development and Aging, 1991, 55, pp. 263–273.

Chapter 4 – Applications of the Derivative

Exercises

Perform the following analysis of the model described above.

1. Graph $f(t)$, $f'(t)$, $f''(t)$ for $0 \leq t \leq 20$.
2. Approximately how much did a rat weigh after 7 days?
3. Approximately when did a rat's weight reach 27 grams?
4. Approximately how fast was a rat gaining weight after 4 days?
5. Approximately when was a rat gaining weight at the rate of 2 grams per day?
6. Approximately when was the rate gaining weight at the fastest rate?

4.6 Analysis of a Botanical Study

The height of the topical bunch-grass elephant millet, t days after mowing, is

$$f(t) = -3.14 + .142t - .0016t^2 + .0000079t^3 - .0000000133t^4$$

meters.[3]

Exercises

Perform the following analysis of the mathematical model described above.

1. Graph $f(t)$, $f'(t)$, $f''(t)$ for $0 \leq t \leq 200$
2. How tall was the grass after 160 days?
3. When was the grass 1.75 meters high?
4. How fast was the grass growing after 80 days?

---

[3] Woodward and Prine, *Crop Quality and Utilization*, Crop Science, 33, 1993, pp. 818–824.

5. When was the grass growing at the rate of .03 meters per day?
6. When was the grass growing at the slowest rate?
7. During the period from the 100th to the 240th day, when was the grass growing at the fastest rate?

4.7 Analysis of a Medical Experiment

The relationship between the area of the pupil of the eye and the intensity of light was analyzed by B. H. Crawford,[4] who concluded that the area of the pupil is

$$f(x) = \frac{160x^{-.4} + 94.8}{4x^{-.4} + 15.8}$$

square millimeters when x units of light are entering the eye per unit time.

Exercises

Perform the following analysis of the mathematical model described above.

1. Graph $f(x)$, $f'(x)$, $f''(x)$ for $0 \leq x \leq 15$.
2. How large is the pupil when 5 units of light are entering the eye per unit time?
3. When 4 units of light are entering the eye per unit time, what is the rate of change of pupil size with respect to a unit change in light intensity?
4. For what light intensity is the pupil size 9 square millimeters?
5. For what light intensity is the rate of change of pupil size with respect to a unit change in light intensity approximately −.2?

---

[4] B. H. Crawford, *The dependence of pupil size upon the external light stimulus under static and variable conditions*, Proc. Royal Society, Series B, 121 (1937), pp. 376–395.

4.8 Mathematical Model of an Illness

A patient's temperature (in degrees Fahrenheit) t hours after contracting an illness is given by

$$T(t) = -.0008t^3 + .0288t^2 + 98.6, \quad 0 \le t \le 36$$

Exercises

Perform the following analysis of the model of the patient's temperature described above.

1. Graph the function $T(t)$.
2. Describe what the graph shows.
3. What is the patient's temperature at time 0?
4. What is the patient's temperature at time 12 hours?
5. Does the mathematical model apply outside the domain $0 \le t \le 36$? Explain your answer.
6. By examining the graph, determine the time at which the maximum temperature occurs.
7. What is the maximum temperature which the patient experiences during the illness?

4.9 Analysis of Coffee Consumption

Coffee consumption in the United States is greater on a per capita basis than anywhere else in the world. However, due to price fluctuations of coffee beans and worries over the health effects of caffeine, coffee consumption has varied considerably over the years. According to data published in the Wall Street Journal, the number of cups y consumed daily per adult in year x (with 1955 corresponding to $x = 0$) is given by the mathematical model:

$$y = 2.76775 + 0.0847943x - 0.00832058x^2 + 0.000144017x^3$$

Exercises

Perform the following analysis of the mathematical model of coffee consumption.

1. Graph y as a function of x to show daily coffee consumption from 1955 through 1994.

2. Determine when coffee consumption was least during this period (to the closest month). What was the daily coffee consumption at that time?

3. Determine when coffee consumption was greatest during this period (to the closest month). What was the daily coffee consumption at that time?

4. At what rate was coffee consumption changing in 1990?

5. When was the rate of change of coffee consumption the most (to the nearest month)?

4.10 Analysis of the Sales of Cough and Cold Medicines

The total amount T (in millions of dollars per week) spent on cough and cold remedies x weeks after the beginning of the cough and cold season (September 27) is given by the model

$$T = 11.25 + 0.9597x - 0.5039x^2 - 0.04133x^3 + 0.0007916x^4$$

Exercises

Perform the following analysis of the above mathematical model.

1. Graph T as a function of x to show amounts spent on cough and cold remedies for the 28 week cough and cold season.

2. Describe in words what the graph portrays.

3. Is the mathematical model realistic? Explain your answer.

4. Determine during which week sales are at a maximum. What is the amount of sales for that week?

5. When is the rate of change of sales greatest?

6. When is the rate of change of sales least?

Techniques of Differentiation

5.1 The Rate of Change of the Election Function

The function

$$f(x) = \frac{x^3}{x^3 + (1-x)^3}, \quad 0 \leq x \leq 1$$

estimates the Democratic proportion of the House of Representatives if x is the proportion of the popular vote for Democrats.

Exercises

Perform the following analysis of the mathematical model of elections given above.

1. Graph this function.
2. Analyze the graph to estimate the effect of an additional 1% Democratic vote from 54% on the composition of the House of Representatives.
3. Calculate the derivative of $f(x)$.
4. Use the derivative to estimate the effect of an additional 1% Democratic vote from 54% on the composition of the House of Representatives.
5. Compare the results of 2. and 4.

5.2 Calculus and Technology Complement Each Other, I

In sketching graphs of functions using MAPLE, it often helps to calculate the derivatives using calculus and then using the graphs of the derivatives to determine relative extreme points and inflection points. This interplay of calculus and MAPLE is illustrated by the following exercises.

Exercises

Sketch the graphs of the following functions using both the approach of your text and by using MAPLE. Be sure to determine all zeros, asymptotes, relative extreme points, and inflection points. Carry out approximations to one decimal place.

1. $f(x) = \dfrac{x^2 + 1}{x^2 - 1}$

2. $f(x) = \dfrac{x}{x^2 + 1}$

3. $f(x) = \dfrac{1}{(x-1)(x+2)}$

4. $f(x) = x^4 + 3x - 1$

5.3 Calculus and Technology Complement Each Other, II

It is possible to be misled if you do an analysis by just looking at the graph of a function, without using any mathematical facts about the function. The following exercises show the problems which can arise if you don't properly set the graph parameters. And the values needed for properly setting the parameters are derived from calculus.

Exercises

Let $f(x) = x^3(1-x)^4$.

1. Make a table of the values of $f(x)$ for $x = -10, -9, ..., 9, 10$.

2. The function values increase rapidly for x away from 0. To get an overall picture of the graph, you need a wide range of y-values. To this end, graph $f(x)$ using the following windows:

    (a) $-2 \leq x \leq 3$, $-100 \leq y \leq 100$
    (b) $-1 \leq x \leq 2$, $-10 \leq y \leq 10$
    (c) $-1 \leq x \leq 2$, $-1 \leq y \leq 1$
    (d) $-1 \leq x \leq 2$, $-.1 \leq y \leq .1$
    (e) $-1 \leq x \leq 2$, $-.01 \leq y \leq .01$

3. At what range of y-values did the graph give a hint that something might be happening between $x = 0$ and $x = 1$?

4. Describe what appears to be happening in graph 2e.

5. Determine the derivative $f'(x)$ using the differentiation techniques of this section. Simplify the derivative. Use the formula for the derivative to explain why further magnification of the y-axis will not reveal addition "bumps" in the graph of $f(x)$.

5.4 Differentiation of Products and Quotients

Exercises

1. $\dfrac{d}{dx}\left[(x^2-3)(x^3+3x-1)^4\right]$

2. $\dfrac{d}{dx}\left[\left(\dfrac{x^2-1}{x+3}\right)^4(3x-1)\right]$

3. $\dfrac{d}{dx}\left[\dfrac{x^2+(1-x)^2}{3x^4+4x^3-2x^2+3}\right]$

4. $\dfrac{d}{dx}\left[(x-1)(x^2+4)(x-3)^2\right]$

5. Evaluate $\dfrac{d}{dx}\left[\dfrac{x^2}{x^2+1}\right]$ at $x=3$.

6. Evaluate $\dfrac{d}{dx}\left[\dfrac{1}{(x^2+1)^5}\right]$ at $x=0$.

7. Determine the values of x for which $\dfrac{d}{dx}\left[x(1+x^2)^4\right]=0$.

8. Determine the values of x for which $\dfrac{d}{dx}\left[\dfrac{x}{1-x^2}\right]=0$.

9. Determine $\dfrac{d^2}{dx^2}\left[\dfrac{x}{1+x}\right]$.

10. Determine $\dfrac{d^2}{dx^2}\left[\dfrac{x}{(1+x)^2}\right]$.

5.5 The Chain Rule

MAPLE "knows" about the chain rule. If $u=g(x)$, then the derivative of $f(u)$ with respect to x can be specified in MAPLE using the command:
`diff(f,x);`

The program responds by displaying the derivative of the composite function $f(g(x))$ with respect to x.

Exercises

Calculate $\dfrac{df}{dx}$ where:

1. $f(u)=\sqrt{u^2+1},\quad u=\dfrac{1}{x^2+1}$

2. $f(u)=u^3+3u+\dfrac{4}{u^2},\quad u=\dfrac{x}{x+1}$

3. $f(u)=\left(u^3+1\right)^4\left(u^2-1\right)^3,\quad u=\sqrt{x^2+1}$

4. $f(u) = \dfrac{u^2-1}{u^2+1}, \quad u = \sqrt{x + \dfrac{1}{x}}$

5.6 Plotting Implicitly-Defined Functions

As we mentioned in Chapter 1, the `solve` command can only plot equations written in the form $y = $ [an expression in x]. In order to plot equations not in this form generally requires the command `implicitplot`, which uses more sophisticated and laborious calculations to determine the points to plot. For example, to plot the equation

$$x^2 + x - y^3 = 1$$

with the variables in the ranges $-1 \leq x \leq 1$, $-1 \leq y \leq 1$, you would use the command:

`with(plots): implicitplot[x^2+x-y^3=1,x=-1..1,y=-1..1]`

Exercises

Plot the following implicitly-defined functions:

1. $x^2 + y^2 = 4, \quad -2 \leq x, y \leq 2$
2. $x^2 + \dfrac{y^2}{4} = 1, \quad -2 \leq x, y \leq 2$
3. $x^4 + 2x^2y^2 + y^4 = 4x^2 - 4y^2, \quad -2 \leq x, y \leq 2$
4. $xy + y^2 = 14, \quad -5 \leq x, y \leq 5$

The Exponential and Natural Logarithm Functions

6.1 Calculating With Exponential Functions

The general exponential function $f(x) = a^x$ may be defined in MAPLE using the command:
`f := x -> a^x;`

The number e is a constant which is built into MAPLE and is denoted `E`. The natural exponential function e^x is likewise built into MAPLE and is denoted `exp(x)`.

The following exercises provide some practice in entering and using exponential functions.

Exercises

For each of the following exponential functions $f(X)$:

    a.   Enter the function in MAPLE.

    b.   Determine the function values $f(1)$, $f(\frac{1}{2})$, $f(4)$, $f(-1)$.

1. $f(X) = 3^X$
2. $f(X) = 2^{-X}$
3. $f(X) = e^{X/2}$

4. $f(X) = e^{-X}$
5. $f(X) = Xe^{-X}$
6. $f(X) = \dfrac{e^X}{X}$
7. $f(X) = 3(1 - e^{-X})$
8. $f(X) = 8e^{2X}$

6.2 Slopes of Exponential Functions

Exponential functions are just another sort of function. Once you define an exponential function, MAPLE may be used to estimate numerically the values of their derivatives at specific points. (See the discussion in Chapter 3.)

Exercises

Calculate the following derivatives.

1. $\frac{d}{dx}[2.5^x]|_{x=0}$
2. $\frac{d}{dx}[2.5^x]|_{x=3}$
3. $\frac{d}{dx}[0.7^x]|_{x=0}$
4. $\frac{d}{dx}[0.7^x]|_{x=-1}$
5. $\frac{d}{dx}[10^x]|_{x=0}$
6. $\frac{d}{dx}[10^x]|_{x=-2}$
7. $\frac{d}{dx}[5.4^x]|_{x=0}$
8. $\frac{d}{dx}[5.4^x]|_{x=1}$

Calculate the derivatives of the following functions at $X = 1$.

9. $f(X) = 3^X$

10. $f(X) = 2^{-X}$

11. $f(X) = e^{X/2}$

12. $f(X) = e^{-X}$

13. $f(X) = Xe^{-X}$

14. $f(X) = \dfrac{e^X}{X}$

15. $f(X) = 3(1 - e^{-X})$

16. $f(X) = 8e^{2X}$

6.3 Graphs of Exponential Functions

You may define functions which involve exponential functions and plot them just as you would any other function.

For example, you would define the function
$$f(x) = xe^{-3x}$$
using the command:
`f := X -> X*e^(-3*X);`
You could then display a portion of its graph using a command of the form:
`plot(f,0..10);`

Exercises

Sketch the graphs of the following functions which are of types which will be used in various applications of the exponential function. Experiment to determine a suitable range which shows the interesting properties of the graph. Determine all asymptotes.

1. $f(X) = 3^X$

2. $f(X) = 2^{-X}$
3. $f(X) = e^{X/2}$
4. $f(X) = e^{-X}$
5. $f(X) = Xe^{-X}$
6. $f(X) = \dfrac{e^X}{X}$
7. $f(X) = 3(1 - e^{-X})$
8. $f(X) = 8e^{2X}$
9. $f(X) = 1000e^{-.07X}$
10. $f(X) = 50e^{3X}$
11. $f(X) = 300(1 - e^{-3X})$
12. $f(X) = e^{-X} + 4$
13. $f(X) = e^{-X^2}$
14. $f(X) = \dfrac{10}{1 + e^{-X}}$

6.4 Maxima and Minima Involving Exponential Functions

Let us now apply to exponential and related functions the techniques introduced earlier to determine relative extreme points, either graphically or by calculating the first and second derivatives.

Exercises

Graphically determine the relative extreme points of the following functions. Carry out your calculations to one decimal place accuracy. Then repeat the analysis calculating the first and second derivatives.

1. $f(x) = xe^{-3x}$

2. $f(x) = \dfrac{1}{1 + e^x}$

3. $f(x) = -x^2 + 3x + e^{-x}$

4. $f(x) = e^{-x} - 10e^{-2x} + 3e^{-3x}$

6.5 Normal Curves

In probability and statistics, it is necessary to consider random variables whose probability distributions are given by the functions

$$f_a(x) = \dfrac{1}{\sqrt{2\pi}} e^{-\frac{1}{2}(x/a)^2}, \quad a > 0$$

The graphs of these functions are called **normal curves**.

Exercises

Solve the following exercises concerning normal curves.

1. Graph the normal curves for $a = 1, 2, 5, 10$.

2. Explain what happens to the normal curves as the value of a increases.

3. Determine the relative extreme points and inflection points of the normal curve with $a = 1$. Carry out your calculations to two decimal places.

6.6 Calculating With the Natural Logarithm Function

In MAPLE, the natural logarithm function is denoted `ln(x)` and the logarithm to base 10 is denoted `log10(x)`.

You work with these functions exactly as you do with any other function, such as the natural exponential function.

Exercises

For each of the following functions:

    a.   Define the function in MAPLE.

    b.   Calculate the function values $f(2)$, $f(10)$.

    c.   Display the graph of the function.

    d.   From the graph determine the domain of the function.

1. $f(X) = \ln 2X$
2. $f(X) = \log(X+1)$
3. $f(X) = X \ln X$
4. $f(X) = \ln(X^2 + 1)$
5. $f(X) = \dfrac{\log X}{X}$
6. $f(X) = \ln(e^X)$
7. $f(X) = \dfrac{1}{X} + \ln X$

6.7 Graphs Involving the Natural Logarithm Function

The natural logarithm function may be used to define other functions, such as the function

$$f(x) = x(\ln x)^2$$

In the following exercise set, we explore the extreme points of some functions obtained in this way.

Exercises

Graph the following functions. Determine their relative extreme points to one decimal place accuracy.

1. $f(x) = x^3 + 10 \ln x$
2. $f(x) = (\ln x)^2$
3. $f(x) = \dfrac{\ln x}{x^2 + 1}$
4. $f(x) = x^2 \ln x$

6.8 Solving Equations Involving Exponential and Logarithmic Functions

You may solve equations involving exponential and logarithmic functions by graphing both sides as functions and determining the intersection points of the graphs or by using the command `fsolve`.

In solving equations graphically, it may require some experimentation to arrive at a set of graph parameters which displays the intersection points.

Applied Calculus Using MAPLE V

Exercises

Solve the following equations both graphically and using `fsolve`. Determine the solutions to two decimal places accuracy.

1. $e^{2x} = 3x + 2$
2. $5e^{-2x} = 2x^2 + 1$
3. $\ln^3 x = x^2 - 10$
4. $e^{-x^2} = \ln x$

6.9 Testing the Laws of Logarithms

Here is a graphical approach to testing the laws of logarithms. While not an actual proof, the following arguments can provide strong evidence that the laws of logarithms actually hold.

Exercises

1. Graph on the same coordinate system the functions $f(x) = \ln x + \ln a$ and $g(x) = \ln(ax)$ for at least 10 positive values of a. What do you observe? What conclusion does the data suggest?

2. Graph on the same coordinate system the functions $f(x) = \ln x - \ln a$ and $g(x) = \ln\left(\dfrac{x}{a}\right)$ for at least 10 positive values of a. What do you observe? What conclusion does the data suggest?

3. Graph on the same coordinate system the functions $f(x) = a \ln x$ and $g(x) = \ln(x^a)$ for at least 10 positive values of a. What do you observe? What conclusion does the data suggest?

Applications of the Exponential and Natural Logarithm Functions

7.1 Internal Rate of Return of an Investment

Suppose that an investment of $2,000 yields payments of $2,000 in 3 years, $1,000 in 4 years and $1,000 in 5 years. Thereafter, the investment is worthless. What constant rate of return r would the investment need to produce in order to yield the payments specified. The number r is called the **internal rate of return** of the investment. We can consider the investment as consisting of three parts, one part yielding each payment. The sum of the present values of the three parts must total $2,000. This yields the equation:

$$2000 = 2000e^{-3r} + 1000e^{-4r} + 1000e^{-5r}$$

Exercises

1. Solve the above equation.

7.2 Elasticity of Demand

Suppose that the number of cars q using a toll road is described by the demand function, where p is toll charged:

$$q = 40,000e^{-.5p} - 2000p + 5000$$

Exercises

Perform the following analysis of the above mathematical model.

1. Graph the demand function.
2. Graph the elasticity of demand $E(p)$.
3. Determine tolls for which the demand is elastic and the tolls for which the demand is inelastic.
4. Graph the revenue function $R(p)$ on the same coordinate system as $E(p)$.
5. Determine the values of p for which $R(p)$ is increasing and those for which $R(p)$ is decreasing.

7.3 The Learning Curve

A common model of learning postulates that the number $N(t)$ of nonsense syllables which can be learned in time t is given by an equation of the form

$$N(t) = M\left(1 - e^{-kt}\right)$$

where M is the maximum number of nonsense syllables which can be learned and k is a constant.

Exercises

Suppose that the maximum number of nonsense syllables which can be learned is 30 and that 10 can be learned in 1 hour.

1. Use a graphical approach to determine the function $y = y(t)$, the number of nonsense syllables which can be learned in time t hours.
2. Graph the function y.

3. Determine the number of nonsense syllables which can be learned in 3 hours.

4. Determine the length of time it takes to learn 20 nonsense syllables.

7.4 The Yield Curve

Each day hundreds of billions of dollars worth of bonds are traded on the world's stock exchanges. At any given moment, bonds of different maturities yield different interest rates. Generally speaking, the longer before a bond is to be refunded, the higher the current interest rate yielded by the bond. If you graph the length of the bond maturity vs its yield, you get a curve called the **yield curve**. Based on data published in Business Week in November 1993, the following equation was derived describing the yield curve: If x is the length of the bond maturity and y is the current interest rate yielded by the bond, then

$$y = 0.062 - 0.031162 e^{-0.2188x}$$

Exercises

Perform the following analyses using the above mathematical model.

1. Graph the yield curve described by the above equation.

2. What is the asymptote on the yield curve? What is the applied significance of the asymptote?

3. What is the yield on a bond which has 8 years left before maturity?

4. How long must the maturity be in order for the yield to be 5.95%?

7.5 Spread of an Epidemic

Suppose that an epidemic of flu spreads among students at a certain college according to the logistic equation

$$N = \frac{4000}{1 + 50e^{-t}}$$

where N denotes the number who have been infected by time t days.

Exercises

Perform the following analyses on the above mathematical model.

1. Graph the function N.
2. What are the asymptotes of the graph?
3. At what rate is the epidemic spreading 5 days after it begins?
4. When is the epidemic spreading fastest? (To answer this, it may be easiest to graph the derivative of N.)

7.6 Analysis of the Effectiveness of an Insect Repellent

Human hands covered with cotton fabrics that had been impregnated with the insect repellent DEPA were inserted for five minutes into a test chamber containing 200 female mosquitos.[5] The function $f(x) = 26.48 - 14.09 \ln x$ gives the number of mosquito bites received when the concentration was x percent.

---

[5] Rao, Prakash, Kumar, Suryanarayana, Bhagwat, Gharia, and Bhavsar *N,N-diethylphenylacetamide in Treated Fabrics as a Repellent Against Aedes aegypti and Culex quinquefasciatus (Deptera: Culiciae)*, Journal of Medical Entomology, vol 28, No. 1, January 1991.

Chapter 7 – Applications of the Exponential and Natural Logarithm Functions

Exercises

Perform the following analyses on the above mathematical model.

1. Graph $f(x)$, $f'(x)$ for $0 \leq x \leq 100$.
2. How many bites were received when the concentration was 3.25%?
3. What concentration resulted in 15 bites?
4. At what rate is the number of bites changing with respect to concentration of DEPA when $x = 2.75$?
5. For what concentration does the rate of change of bites with respect to concentration equal 10 bites per percentage increase in concentration?

7.7 Analysis of the Absorption of a Drug

After a certain drug is taken orally, the amount of the drug in the bloodstream after t hours is $f(t) = 122(e^{-.2t} - e^{-t})$ units.

Exercises

1. Graph $f(t)$, $f'(t)$, $f''(t)$ for $0 \leq t \leq 15$.
2. How many units of the drug are in the bloodstream after 7 hours?
3. At what rate is the level of drug in the bloodstream increasing after 1 hour?
4. Approximately when (while the level is decreasing) is the level of the drug in the bloodstream 20 units?
5. What is the greatest level of drug in the bloodstream and when is this level reached?
6. When is the level of the drug in the bloodstream decreasing the fastest?

7.8 Analysis of the Growth of a Tumor

A cancerous tumor has volume $f(t) = 1.825^3 (1 - 1.6e^{-.4196t})^3$ milliliters after t weeks, with $t > 1$.[6]

Exercises

1. Graph $f(t)$, $f'(t)$, $f''(t)$ for $0 \leq t \leq 15$.
2. Approximately how large is the tumor after 14 weeks?
3. Approximately when will the tumor have a volume of 4 milliliters?
4. Approximately how fast is the tumor growing after 5 weeks?
5. Approximately when is the tumor growing at the rate of .5 milliliters per week?
6. Approximately when is the tumor growing at the fastest rate?

7.9 Growth of a Bacteria Culture With Growth Restrictions

A model incorporating growth restrictions for the number of bacteria in a culture after t days is given by:

$$f(t) = 9000(20 + te^{-.04t})$$

Exercises

1. Graph $f(t)$, $f'(t)$, $f''(t)$ for $0 \leq t \leq 100$.
2. How large is the culture after 20 days?

---

[6] Baker, Goddard, Clark, and Whimster, *Proportion of Necrosis in Transplanted Murine Adenocarcinomas and Its Relationship to Tumor Growth*, Growth, Development and Aging, 1990, 54, pp. 85–93

3. How fast is the culture changing after 100 days?

4. Approximately when are there 230,000 bacteria?

5. Approximately when is the culture growing at the rate of 1500 bacteria per day?

6. When is the size of the culture decreasing the fastest?

7. When is the size of the culture greatest?

The Definite Integral

8.1 Determining Antiderivatives

Symbolic mathematics programs typically have a command for antidifferentiation. In MAPLE, this command is `int`. For example, to calculate an antiderivative of x^2, you would give the command:

`int(x^2,x);`

Note that the second parameter, `x`, indicates the variable for the integration. In response to this command, the program gives the antiderivative:
`1/3 X^2`
Note that the program does not supply the arbitrary constant.

Exercises

to calculate the following antiderivatives:

1. $\int (3x - 4x^6)\,dx$
2. $\int (5 - 3x + 2x^3 + 9x)\,dx$
3. $\int e^{.003x}\,dx$
4. $\int e^{-40x}\,dx$
5. $\int \dfrac{5}{3+2x}\,dx$
6. $\int \dfrac{x}{5x^2-1}\,dx$
7. $\int xe^{-4x}\,dx$
8. $\int xe^{-4x^2}\,dx$
9. $\int (4x+1)e^{-9x-6}\,dx$

10. $\int (x^3 - 4)(x^5 + 1)^4 dx$

8.2 Determining Antiderivatives, II

Suppose that f is a given function and F an antiderivative of f:

$$F(x) = \int f(x)\, dx$$

The function f has an infinite number of antiderivatives, all of which differ from F by a constant. In many problems, a particular antiderivative is specified by a boundary condition which gives the value of F at a particular point. Determining the antiderivative then involves solving an equation, which can be done using `solve`.

Exercises

Use the above approach to determine the antiderivative of f satisfying the given boundary condition.

1. $f(x) = 3.1x - 2.2$, $F(1) = 10.9$
2. $f(x) = x(2.2x^2 + 1)^3$, $F(5.1) = 12.3$
3. $f(x) = 107.9e^{-3.58x}$, $F(100) = 40.1$
4. $f(x) = \dfrac{10.3}{x}$, $F(1) = 4$
5. The most general antiderivative of $f(x) = x^3$ is $F(x) = \frac{1}{3}x^3 + C$, where C is a constant. Graph antiderivatives corresponding to three different values of C on the same coordinate system. Describe the relationship between the graphs. Can you make a conclusion about the graphs of the various antiderivatives of a function?

8.3 Antidifferentiation in Closed Form Is Not Always Possible

Not all simple functions have simple antiderivatives. A number of complications in antidifferentiation can arise. An antiderivative may involve advanced functions, which you may have no knowledge of. Or an antiderivative may not be possible in terms of "standard" functions. In the latter case, MAPLE gives the original problem as its answer. Another possibility is that an antidifferentiation problem may be too complicated for the program. Antidifferentiation is a complicated task and can often require mathematical ingenuity of a type that is difficult to build into a computer program. The following problems should provide some interesting results from MAPLE.

Exercises

Attempt to use MAPLE to determine the following indefinite integrals.

1. $\int \sqrt{1-x^2}\, dx$
2. $\int e^{-x^2} dx$
3. $\int \dfrac{dx}{\sqrt{1-x^2}}$
4. $\int \dfrac{dx}{\sqrt{1-x^3}}$

8.4 Estimation of Integrals Using Riemann Sums

As we have just described, it is possible to estimate definite integrals using Riemann sums. For small numbers of subdivisions, this can be done by hand. However, when the number of subdivisions increases, some form of technology is a necessity. Basically, what is needed is a way to add up sums of numbers which are specified as values of a function. In MAPLE, you can use the function sum. For example, to compute the sum

$$5^2 + 6^2 + 7^2 + \ldots + 100^2$$

you would give the command:
`sum(n^2,n=5..100);`

Exercises

The following exercises provide some practice in calculating Riemann sums. Let $f(x) = 4 - x^2$. Calculate the following Riemann sums for the function $f(x)$ over the interval $-2 \leq x \leq 2$:

1. $n = 10$, right endpoints
2. $n = 50$, right endpoints
3. $n = 100$, right endpoints
4. $n = 500$, right endpoints
5. $n = 1000$, right endpoints
6. $n = 10$, left endpoints
7. $n = 50$, left endpoints
8. $n = 100$, left endpoints
9. $n = 500$, left endpoints
10. $n = 1000$, left endpoints
11. $n = 10$, midpoints
12. $n = 50$, midpoints
13. $n = 100$, midpoints
14. $n = 500$, midpoints
15. $n = 1000$, midpoints
16. On the basis of the above data, what is the approximate area bounded by the graph of f and the x-axis? To how many decimal places does the approximation appear to be accurate?

8.5 Calculating Definite Integrals

The `int` command, which allows you to calculate indefinite integrals, may also be used to calculate definite integrals. For example, to calculate the definite integral

$$\int_1^2 x^2 dx$$

you may give the command:
`int(x^2,x=1..2);`
MAPLE responds with the answer:

$$\frac{7}{3}$$

Exercises

Calculate the following definite integrals.

1. $\int_0^3 (3x+1)^5 dx$

2. $\int_1^{20} \sqrt{5x-2}\ dx$

3. $\int_3^5 (x+1)(x-3)^4 dx$

4. $\int_0^1 \frac{x+1}{x+2}\ dx$

5. $\int_{12}^{15} 51e^{-.02t} dt$

6. $\int_0^9 \left[3+e^{-4x}\right]^3 dx$

8.6 Exploring the Fundamental Theorem of Calculus, I

The following sequence of exercises allows you to graphically explore the Fundamental Theorem of Calculus.

Exercises

Let $f(x) = x^3 - 9x^2 + 15x + 20$. Find an antiderivative $F(x)$ of $f(x)$ such that $F(0) = 0$. (Do this without using MAPLE.)

1. Graph $f(x)$ and $F(x)$ using the window $-2 \leq x \leq 8$, $-10 \leq y \leq 100$.

2. Determine an interval $a \leq x \leq b$ on which $f(x)$ has nonnegative values. Describe what happens to the graph of $F(x)$ on this interval.

3. Examine the graph to estimate the values of x at which $f(x)$ has a relative extreme point. Describe what happens to the graph of $F(x)$ at these values of x.

8.7 Exploring the Fundamental Theorem of Calculus, II

The following sequence of exercises allows you to further graphically explore the Fundamental Theorem of Calculus.

Exercises

Let $f(x) = x^3 - 7x^2 + 8x + 10$. Find an antiderivative $F(x)$ of $f(x)$ such that $F(0) = 0$. (Do this without using MAPLE.) The solve the following exercises.

1. Graph $f(x)$ and $F(x)$ using the window $-1 \leq x \leq 10$, $-10 \leq y \leq 50$.

2. By examining the graph, estimate the nonnegative values of x at which $F(x)$ has a relative extreme point. Describe what happens to the graph of $f(x)$ at these values of x.

3. Use the graph of $f(x)$ to explain why $F(x)$ is decreasing over the interval between the relative extreme points found in 2.

8.8 The Area Between Curves

One of the most tedious tasks in calculating the area bounded by curves is calculating the intersection points of the curves. Of course, this can be done using the **solve** or **fsolve** commands and then using the results to set up the necessary integrals. We may then use MAPLE to evaluate the integrals.

Exercises

Use the above approach to calculate the area bounded by the following curves:

1. $y = 5x - 3$, $y = 2x^2 + x - 4$
2. $y = 1 + x$, $y = x^3 - 5x + 1$
3. $y = \dfrac{50}{x+1}$, $y = 3x + 1$, $y = 0$, $x = 0$
4. $y = x^2 + x + 1$, $y = \sqrt{x} + 3$, $x = 0$

8.9 The Volume of a Solid of Revolution

If the area under the graph of

$$y = f(x), \ a \leq x \leq b$$

is revolved about the x-axis, the volume of the resulting solid is given by the integral:

$$\int_a^b \pi [f(x)]^2 dx$$

This integral may be evaluated using MAPLE command. Note that π may be entered as `Pi`. So, for example, the above integral in the case of the function $f(x) = x$ would be entered as:
`int(Pi*x^2,x=a..b);`

Exercises

Estimate the volume of the solids of revolution generated by the revolving the following functions about the x-axis:

1. $y = \dfrac{x}{x+1}, \ 0 \leq x \leq 5$
2. $y = \sqrt{5x-1}, \ 2 \leq x \leq 4$
3. $y = (x^2+1)^6, \ 0 \leq x \leq 4$
4. The area bounded by $y = 2x+5$ and $x^2 - 3x + 5$

8.10 Consumers' Surplus

In the text, the concept of consumer's and producer's surplus is defined. Calculating consumer's and producer's surplus, it is necessary to determine the intersection point of supply and demand curves, a task which may be accomplished using the `solve` or `fsolve` commands. Calculating consumer's and producer's surplus then requires evaluation of an integral, a task which can also be accomplished with MAPLE.

Exercises

Estimate the intersection point of the following supply and demand curves. Then determine the consumers' surplus.

1. Supply curve $p = 500/(x+10)$; demand curve $p = 11 + .04x^2$
2. Supply curve $p = 225/(x+5)$; demand curve $p = 15 + 2.4x^2$

Functions of Several Variables

9.1 Defining and Evaluating Functions of Several Variables

You may define a function of two or more variables in MAPLE in a manner similar to that used to define functions of a single variable. For example, to define the function

$$f(x,y) = xe^{-2xy}$$

you may use the command:

`f := (x,y) -> x*exp(-2*x*y);`

In order to compute the function value $f(2,3)$, you may use the command:

`f(2,3);`

MAPLE responds with the answer:

$$2e^{(-12)}$$

Exercises

Define the following functions in MAPLE and compute the function value at the point given.

1. $f(x,y) = 5x^3 - 34y, \quad (x,y) = (14.1, 12.6)$
2. $f(x,y) = \dfrac{x+y}{x-y}, \quad (x,y) = (-15.3, 101.5)$
3. $f(x,y) = 379x^{1/3}y^{2/3}, \quad (x,y) = (89, 188)$
4. $f(x,y) = 3x^2 - 12xy + 4y^3 + 3y, \quad (x,y) = (-59, 103)$

9.2 Graphing Functions of Two Variables

MAPLE allows you to plot functions in two variables using the command `plot3D`. This command works like the command `plot`, except that it is necessary to specify ranges for both `x` and `y`. Optionally, you may specify a range for `z`.

For example, to display the graph of the function

$$f(x,y) = 3x + 2y, \quad -1 \leq x \leq 2, \; 0 \leq y \leq 3$$

you may use the command:
`plot3D(3*x+2*y,x=-1..2,y=0..3);`
Alternatively, you may define the function `f` as described in the preceding section and use the command:
`plot3D(f(x,y),x=-1..2,y=0..3);`

Exercises

Display the graphs of the following functions of two variables.

1. $x - y - 1, \quad -5 \leq x,y \leq 5$
2. $3x + 2x - 6, \quad -10 \leq x,y \leq 10$
3. $x - 4, \quad -5 \leq x,y \leq 5$
4. $y + 1, \quad -5 \leq x,y \leq 5$
5. $x^2 + y^2, \quad -3 \leq x,y \leq 3$
6. $x - y^2 - 4, \quad 0 \leq x \leq 5, \; -2 \leq y \leq 2$
7. $xy, \quad -2 \leq x,y \leq 2$
8. $x^3 y, \quad -2 \leq x,y \leq 2$
9. $\dfrac{1}{xy}, \quad 0 \leq x,y \leq 10$
10. $\dfrac{x}{y}, \quad 0 \leq x,y \leq 10$

9.3 Graphing Level Curves

Suppose that we are given a function $f(x, y)$ of two variables. A level curve of this function is the graph of an equation

$$f(x, y) = c$$

for a particular constant c. In drawing level curves, it is usually necessary to use the command `implicitplot`.

Exercises

Plot the indicated level curves of the specified function $f(x, y)$.

1. $f(x, y) = 3x + 2y, \quad c = -1, 0, 1, 2$
2. $f(x, y) = xy, \quad c = 0, 1, 2, 3$
3. $f(x) = x^2 + 3y^2, \quad c = 4, 10, 25$
4. $f(x, y) = x - y^2, \quad c = 0, 1, 2, 5$
5. $f(x, y) = \dfrac{x}{y}, \quad c = -2, -1, 0, 1, 2$
6. $f(x, y) = x^3 + y^2, \quad c = 1, 3, 6, 10$

9.4 Calculating Partial Derivatives

In MAPLE, you may calculate partial derivatives using the `diff` command. For example, to calculate

$$\frac{\partial}{\partial x}\left(3x^3 y - \frac{4}{x+y}\right)$$

you could use the command:
```
diff(3*x^3*y-4/(x+y),x);
```

Applied Calculus Using MAPLE V

Note that the last **x** indicates that you are taking the partial derivative with respect to **x**, To calculate:

$$\frac{\partial^2}{\partial x^2}\left(3x^3y - \frac{4}{x+y}\right)$$

you could then give the command:

`diff(",x);`

And to calculate

$$\frac{\partial^2}{\partial x^2}\left(3x^3y - \frac{4}{x+y}\right)\bigg|_{(x,y)=(2,-1)}$$

you may use the **subs** command to substitute **x=2** and **y=-1** in the previous result:

`subs(x=2,y=-1,");`

Exercises

Suppose that:

$$f(x,y) = x^3 y^2 e^{-x^2 y}$$

Calculate the following:

1. $\dfrac{\partial f}{\partial x}$

2. $\dfrac{\partial f}{\partial y}$

3. $\dfrac{\partial^2 f}{\partial x^2}$

4. $\dfrac{\partial^2 f}{\partial y^2}$

5. $\dfrac{\partial^2 f}{\partial x \partial y}$

6. $\dfrac{\partial^2 f}{\partial y \partial x}$

7. $\dfrac{\partial^3 f}{\partial^2 x \partial y}$

8. $\dfrac{\partial^4 f}{\partial x^2 \partial y^2}$

9.5 Solving Optimization Problems in Two Variables

In solving optimization problems in several variables, it is often difficult to solve the system of equations obtained by setting the partial derivatives equal to 0. Moreover, it is often difficult to determine whether a point is a relative maximum or relative minimum. We may solve systems of equations in several variables using **solve** or **fsolve**. Once you determine solutions which are potential extreme points, you can determine whether they are relative maxima, relative minima, or neither by graphing the function in a region near the point.

Exercises

Determine the relative maxima and relative minima of the following functions.

1. $5x^2 + 4y^2 - 13x + 10y + 10$
2. $(3x - 1)^4 + 10y^2 + 6y - 4$
3. $x^4 + 10xy + y^6 - 50$
4. $x^2y + 10x^3y^2 + x + y - 3$

The Trigonometric Functions

10.1 Evaluating Trigonometric Functions

MAPLE provides built-in access to the trigonometric functions $\sin x$, $\cos x$, $\tan x$. They may be used, just like the exponential and logarithmic functions, in expressions and in stored functions. Note, however, that you must always use parentheses in using these functions. That is, you must write `tan(x)` rather than `tan x`. Also, instead of writing $\tan^2 x$, as is customary in mathematics texts, you must write `tan(x)^2`.

Exercises

Enter the following function:

$$f(x) = \sin x \cos^2 x$$

1. Make a list of the function values for $x = 0, .1, .2, ..., 6.3$. (These values of x cover one period of the function since $2\pi \approx 6.28$.)

2. Evaluate the function for $x = 4.38$, 9.12, 10^{-3}.

3. Calculate the derivative of the function at $x = 1.45$.

4. Calculate $\int_0^{2\pi} f(x)dx$.

10.2 Graphing Trigonometric Functions

You may graph a function involving trigonometric functions just as you would any other functions. Note, however, in specifying x range, it is often convenient to use the built-in constant Pi. For example to graph one period of the function sin(x), you could use the command:
`plot(sin(x), x=0..2*Pi);`

Exercises

Graph the following functions for $0 \leq x \leq 2\pi$. Determine all zeros, relative maxima, and relative minima.

1. $\sin x + \cos x$
2. $\tan\left(2x + \frac{\pi}{4}\right)$
3. $\sin^2 x$
4. $\sec x = \dfrac{1}{\cos x}$
5. $\dfrac{x}{\sin x}$
6. $-3\cos(2x)$

10.3 Amplitude and Period of Sine and Cosine Functions

In the following exercises, explore some properties of the functions $f(x) = a\sin(bx)$ and $g(x) = a\cos(bx)$ for constants a and b.

Exercises

1. On a single coordinate system, graph the functions $\sin x$, $\sin 2x$, $\sin 3x$ in the range $-2\pi \leq x \leq 2\pi$. How many repetitions of the each graph are there?

2. On a single coordinate system, graph the functions $\sin x$, $\sin(x/2)$, $\sin(x/3)$ in the range $-6\pi \leq x \leq 6\pi$. How many repetitions of the each graph are there?

3. A number c is called a **period** of the function f provided that the graph of f repeats every c units along the x-axis; that is,
$$f(x+c) = f(x)$$
for all x. Based on your observations in 1. and 2. what are the periods of the functions $\sin x$, $\sin 2x$, $\sin 3x$ and $\sin x$, $\sin(x/2)$, $\sin(x/3)$?

4. Based on your observations, can you guess the period of the function $\sin bx$?

5. Repeat 1-4, substituting the cosine function for the sine.

6. On a single coordinate system, graph the functions $\sin x$, $2\sin x$, $\frac{1}{2}\sin x$ in the range $-2\pi \leq x \leq 2\pi$. How are the graphs related to one another?

7. Repeat 6. replacing the sine with the cosine.

8. On the basis of 6. and 7. can you make a guess as to the geometric significance of the constant a in the graphs of the functions $f(x) = a\sin(bx)$ and $g(x) = a\cos(bx)$? The number a is called the **amplitude**.

10.4 Analysis of Blood Pressure Readings

The pressure of the blood being pumped by the heart can be described by a model involving the cosine function. Suppose that the blood pressure at time t minutes is given by the function:
$$f(t) = 100 + 30\cos 150\pi t$$

Exercises

Perform the following analysis of the above mathematical model.

1. Draw the graph of $f(t)$. Choose a suitable interval of time over which to graph the function so that you may easily read the graph.

2. Describe in words the behavior of the blood pressure.

3. How many times per minute is the heart beating? (One heartbeat corresponds to one cycle of blood pressure.)

4. What are the maximum and minimum blood pressure readings?

10.5 The Lotka-Volterra Predator-Prey Model

The Lotka-Volterra Predator-Prey model is a mathematical model which describes the interaction between a predator species and its prey. As a particular example of this model, suppose that the number of predators in a particular geographic region at time t is given by the function

$$N(t) = 5000 + 2000 \cos\left(\frac{\pi}{18}t\right)$$

where t is measured in months after June 1, 1996.

Exercises

Perform the following analysis of the above mathematical model.

1. Graph the number of predators versus time. Choose an appropriate range for t so that all important geometric facts about the graph are displayed.

2. What is the period of the graph?

3. During a single period, beginning June 1, 1990, when is the number of predators increasing?

4. During a single period, beginning June 1, 1990, when is the number of predators decreasing?

5. What is the maximum number of predators? When is the maximum achieved in the first period beginning Jun 1, 1990?

6. What is the minimum number of predators? When is the minimum achieved in the first period beginning Jun 1, 1990?

10.6 Number of Hours of Daylight

The number of hours of daylight D depends on the latitude and the day t of the year and is given by the function

$$D(t) = 12 + A\sin\left[\frac{2\pi}{365}(t-80)\right]$$

where A depends only on the latitude (and not on t). For latitude 30° the value of A is approximately 2.3.

Exercises

1. Graph the function $D(t)$ for latitude 30°.

2. When is the number of hours of daylight greatest? What is the length of that day?

3. When is the number of hours of daylight least? What is the length of that day?

4. When is the length of the day 11 hours?

5. When is the length of the day 13 hours?

6. When is the number of hours of daylight increasing at a rate of 2 minutes per day?

7. When is the number of hours of daylight decreasing at a rate of 2 minutes per day?

8. When is the number of hours of daylight equal to the number of hours of darkness? (These times are called **equinoxes**.)

10.7 Graphs of Functions Formed From Trigonometric Functions

In this set of exercises, we combine the trigonometric functions with various other functions and examine their graphs for periodicity.

Exercises

Graph the following functions. Determine if the graphs are periodic and if so determine the period.

1. $f(x) = \tan 3x$
2. $f(x) = \sec 2x$
3. $f(x) = \tan\left(\dfrac{x}{2} + \dfrac{\pi}{8}\right)$
4. $f(x) = \sec \dfrac{x}{4}$
5. $f(x) = 3\sin 2x + \cos x$
6. $f(x) = 4\sin x + 5\sin \dfrac{x}{3}$
7. $f(x) = x \sin x$
8. $f(x) = \dfrac{\sin x}{x}$
9. $f(x) = e^{-x} \sin x$
10. $f(x) = x + \sin x$

Techniques of Integration

11.1 More Antidifferentiation

MAPLE has built-in all of the techniques of integration covered in your text, especially integration by substitute and integration by parts. The following exercises illustrate how MAPLE handles indefinite integrals requiring these techniques.

Exercises

Use MAPLE to calculate the following antiderivatives:

1. $\int (3x^2 + 4)(x^3 + 4x)^{-1/2} e^{-\sqrt{x^3+4x}} \, dx$
2. $\int \cot x \cdot \ln \sin x \, dx$
3. $\int \sin^2(\cos x) \sin x \, dx$
4. $\int (3x - 1)^3 (4x + 1)^7 (3x + 1)^9 \, dx$

11.2 Still More Antidifferentiation

Use MAPLE to calculate the following antiderivatives. Some of them can be calculated using the methods of this section, although with more computation than required by the examples or exercises. Others can be calculated using other methods which are not covered in this text, but which are "known" to MAPLE program.

Exercises

1. $\int x^5 e^{-4x} dx$
2. $\int (3x-1)^5 \ln x \; dx$
3. $\int \sin 5x \cos x \; dx$
4. $\int x \tan x \; dx$
5. $\int \sin^3 x \; dx$
6. $\int e^x \sin x \; dx$
7. $\int \sin^4 x \cos^6 x \; dx$
8. $\int \dfrac{dx}{(x-1)(x-2)(2x+1)}$
9. $\int (2x-1)^4 \ln x \; dx$
10. $\int \dfrac{dx}{3x^2 - 2x - 1}$

11.3 The Trapezoid Rule

Suppose that the function $f(x)$ is defined on the interval $A \leq x \leq B$ and that N is a positive integer. Then the trapezoid rule for approximating the integral $\int_A^B f(x) dx$ asserts that:

$$\int_A^B f(x) dx \approx$$

$$\frac{1}{2}[f(A) + 2f(A + \Delta x) + 2f(A + 2\Delta x) +$$
$$\ldots + 2f(A + (N-1)\Delta x) + f(B)]\Delta x$$

where $\Delta x = (B - A)/N$. In order to obtain accurate results, it is necessary to use this formula with relatively large values of N. This requires a fair

Applied Calculus Using MAPLE V

amount of calculation. The above expression may be calculated in MAPLE using the commands:

```
deltax := (B-A)/N;
0.5*(f(A)+sum(2*f(A+n*deltax),n=1..N-1)+f(B))*deltax;
```

Exercises

1. Test the above commands using the data $f(x) = x$, $A = 0$, $B = 1$, $N = 5$. Compare the result of the commands described above with the result obtained by carrying out the trapezoid rule by hand.

Estimate the following integrals using the trapezoid rule with $N = 10, 25, 100, 500$. Compare the program running times as N is increased. How much does running time increase if N is doubled?

2. $\int_0^1 x^3 dx$

3. $\int_0^{10} x^2 e^{-x} dx$

4. $\int_2^{100} \frac{1}{\ln x} dx$

5. $\int_1^3 e^{-x^2} dx$

11.4 Simpson's Rule

Suppose that the function $f(x)$ is defined on the interval $A \leq x \leq B$ and that N is a positive, even integer. Then Simpson's rule for approximating the integral $\int_A^B f(x)dx$ asserts that:

$$\int_A^B f(x)dx \approx$$
$$\frac{1}{3}[f(A) + 4f(A+\Delta x) + 2f(A+2\Delta x) + ...$$
$$+ 2f(A+(N-2)\Delta x) + 4f(A+(N-1)\Delta x) + f(B)]\Delta x$$

where $\Delta x = (B-A)/N$. In order to obtain accurate results, it is necessary to use this formula with relatively large values of N. This requires a fair amount of calculation. All of the calculation may be carried out using the following MAPLE commands:

```
deltax := (B-A)/N;
evenodd := proc(n) if n/2 = trunc((n+1)/2)) then 1
           else 2 fi end;

(1/3)*(f(A)+4*f(A+deltax)+
       2*sum(evenodd(n)*f(A+n*deltax),n=2..N-1)+f(B))*deltax;
```

The function `evenodd(n)` equals 1 if n is odd and 2 if n is even.

Exercises

1. Test the above commands using the data $f(x) = x$, $A = 0$, $B = 1$, $N = 10$. Compare the result of the program with the result obtained by carrying out the trapezoid rule by hand.

Estimate the following integrals using Simpson's rule with $N = 10, 100, 500$. Compare the program running times as N is increased. How much does running time increase if N is doubled?

2. $\int_0^1 x^3 dx$

3. $\int_0^{10} x^2 e^{-x} dx$

4. $\int_2^{100} \frac{1}{\ln x} dx$

5. $\int_1^3 e^{-x^2} dx$

11.5 Comparing the Accuracy of the Trapezoid and Simpson's rules

In this set of exercises, we explore the accuracy produced by the trapezoid and Simpson's rule for a given value of N.

Exercises

Approximate the following integrals using (a) the trapezoid rule for $N = 100, 500, 1000$, (b) Simpson's rule for $N = 100, 500, 1000$. In each case, estimate the number of decimal places of accuracy provided by each approximation. Based on the data collected, what can you say about the relative accuracy of the two rules?

1. $\int_0^2 x\, dx$

2. $\int_1^3 \frac{1}{x}\, dx$

3. $\int_0^{\pi/4} \sin x\, dx$

4. $\int_{-1}^1 e^{-x^2}\, dx$

11.6 Approximating Definite Integrals With Specified Accuracy

We may estimate the error in the trapezoid and Simpson's rule as follows.

1. The error in the trapezoid rule is at most
$$\frac{M(B-A)^3}{12N^2}$$
where M is a number such that:
$$|f''(X)| \leq M \text{ for } A \leq X \leq B$$

2. The error in Simpson's rule is at most
$$\frac{K(B-A)^5}{2880N^2}$$
where K is a number such that:
$$|f^{(4)}(X)| \leq K \text{ for } A \leq X \leq B$$

You may obtain values for M and K by examining the graphs of $f''(X)$ and $f^{(4)}(X)$, respectively. Once you have values for M and K, you may determine the value of N which gives a maximum specified error in each case.

Exercises

Use the approach described above to determine the number N of subintervals required by (a) the trapezoid rule and (b) Simpson's rule to approximate the following integrals to with an error of 10^{-5}:

1. $\int_0^1 (5X-3)^7 e^{-X} dX$

2. $\int_0^{\pi/4} \sqrt{\sin X} \, dX$

3. $\int_1^{100} \ln^5 X \, dX$

4. $\int_0^1 \frac{X}{X+1} \, dX$

11.7 Calculating Improper Integrals

You may estimate the value of the improper integral

$$\int_A^\infty f(X)dX$$

by using the command:
`int(f(x),x=A..infinity);`

Exercises

Use the above approach to estimate the values of the following improper integrals.

1. $\int_1^\infty 41e^{-.03X}dX$

2. $\int_1^\infty (3X^{-2}+4X^{-3})^4 dX$

3. $\int_0^\infty X^3 e^{-5X} dX$

4. $\int_1^\infty \frac{\ln X}{X^3} dX$

Calculus and Probability

12.1 The Chi-Square Probability Density Function

The chi-square probability distribution is used extensively in statistics work. It depends on a parameter d, which indicates the number of degrees of freedom (defined in a statistics course). The density function for the chi-squared distribution is given by the function:

$$f_d(x) = \frac{1}{2^{d/2}\Gamma(d/2)} e^{-x/2} x^{d/2-1}$$

Here $x \geq 0$ and Γ denotes the gamma function (which corresponds to the MAPLE function **gamma**). In MAPLE, this function can be defined using the formula:

```
f := (x,d) -> 1/2^(d/2)*gamma(d/2)exp(-x/2)*x^(d/2-1);
```

Exercises

1. Determine the probability distribution function corresponding to 2 degrees of freedom.

2. Graph the density function for a chi-square random variable with 2 degrees of freedom.

3. Determine the probability that a chi-square random variable with 2 degrees of freedom is between 0 and 4; is greater than or equal to 8.

4. Determine the probability distribution function corresponding to 6 degrees of freedom.

5. Graph the density function for a chi-square random variable with 6 degrees of freedom.

6. Determine the probability that a chi-square random variable with 6 degrees of freedom is less than or equal to 5; greater than or equal to 10.

12.2 Calculating Expected Value and Variance

Refer to the preceding section for the formula for the chi-square distribution function.

Exercises

1. Calculate the expected value of a chi-square random variable with 2 degrees of freedom.

2. Calculate the variance of a chi-square random variable with 2 degrees of freedom.

3. Calculate the expected value of a chi-square random variable with 6 degrees of freedom.

4. Calculate the variance of a chi-square random variable with 6 degrees of freedom.

Taylor Polynomials and Infinite Series

13.1 Approximation by Taylor Polynomials

In MAPLE, you may obtain the nth Taylor polynomial of `f(x)` about `x=a` using the command:
`taylor(f(x),x=a, n);`
MAPLE uses its ability to calculate derivatives to respond with the desired Taylor polynomial.

Exercises

1. Determine the first four Taylor polynomials about $x = 0$ of the rational function:
$$f(x) = \frac{1}{1 + x + x^2}$$

2. Graph each Taylor polynomial and $f(x)$ on the same coordinate system. By examining the graphs, determine how closely each Taylor polynomial approximates $f(x)$ for x in the interval $-1 \leq x \leq 1$.

3. Use a graphical approach to determining upper bounds on the first 5 derivatives of $f(x)$ for $-1 \leq x \leq 1$. Use the bounds to estimate the error in the Taylor polynomial approximations using the result in the text.

4. Compare the results of 3. and the results from 2. Which method yields better results? Why?

13.2 Bernoulli Numbers

Determine the first ten Taylor polynomials for

$$f(x) = \frac{x}{e^x - 1}$$

The coefficients of the Taylor polynomials are an interesting set of fractions called **Bernoulli numbers**. These numbers have many important properties which are derived in more advanced mathematics courses.

13.3 The Newton-Raphson Method

You may use MAPLE to perform the calculations of the Newton-Raphson method.

Exercises

Use the Newton-Raphson method to determine all zeros of the following functions to 5 decimal-place accuracy.

1. $x^3 + x^2 - 4x + 1$
2. $-x^3 + 3x^2 + 36x + 25$
3. $x^2 - \sin 2x$
4. $e^{-.1x} - x^3$

13.4 Two Interesting Infinite Series

In this set of exercises, we explore two infinite series, one convergent and one divergent.

Chapter 13 – Taylor Polynomials and Infinite Series

Exercises

1. It can be shown that the series
$$1 - \frac{1}{3} + \frac{1}{5} - \frac{1}{7} + \dots$$
converges and has the sum $\frac{\pi}{4}$, a fact proved by Gregory in 1607. Use a symbolic math program to calculate the partial sum of the first 100 terms; 1,000 terms; 10,000 terms; 100,000 terms. Estimate how many terms are necessary to use to obtain a given number of decimal places of $\frac{\pi}{4}$.

2. The series of 1. can be written as
$$\left(\frac{1}{1} - \frac{1}{3}\right) + \left(\frac{1}{5} - \frac{1}{7}\right) + \left(\frac{1}{9} - \frac{1}{11}\right) + \dots$$
$$= \frac{2}{1 \cdot 3} + \frac{2}{5 \cdot 7} + \frac{2}{9 \cdot 11} + \dots$$

   Calculate the partial sum of the first 10 terms of this series; 100 terms; 1,000 terms. Compare each partial sum with the sum of the series. Estimate the number of terms required to obtain a given number of decimal places of $\frac{\pi}{4}$. (This example shows that by slightly altering an infinite series, we can make it converge to its sum much faster.)

3. Consider the **harmonic series**:
$$1 + \frac{1}{2} + \frac{1}{3} + \frac{1}{4} + \dots$$

   Calculate the sum of the first 1,000 terms; 10,000 terms; 100,000 terms; 1,000,000 terms. What conclusion do these results suggest?